Dimensions Math®
Teacher's Guide 2B

Authors and Reviewers

Cassandra Turner

Allison Coates

Jenny Kempe

Bill Jackson

Tricia Salerno

Singapore Math Inc.

Published by Singapore Math Inc.

19535 SW 129th Avenue
Tualatin, OR 97062
www.singaporemath.com

Dimensions Math® Teacher's Guide 2B
ISBN 978-1-947226-35-7

First published 2018
Reprinted 2019, 2020, 2021

Copyright © 2017 by Singapore Math Inc.
All rights reserved. This book or any portion thereof may not be reproduced or used in any manner whatsoever without the express written permission of the publisher.

Printed in China

Acknowledgments

Editing by the Singapore Math Inc. team.
Design and illustration by Cameron Wray with Carli Bartlett.

Contents

Chapter		Lesson	Page
Chapter 8 **Mental Calculation**		Teaching Notes	1
		Chapter Opener	5
	1	Adding Ones Mentally	6
	2	Adding Tens Mentally	8
	3	Making 100	10
	4	Adding 97, 98, or 99	12
	5	Practice A	14
	6	Subtracting Ones Mentally	16
	7	Subtracting Tens Mentally	18
	8	Subtracting 97, 98, or 99	20
	9	Practice B	22
	10	Practice C	24
		Workbook Pages	26
Chapter 9 **Multiplication and Division of 3 and 4**		Teaching Notes	33
		Chapter Opener	37
	1	The Multiplication Table of 3	38
	2	Multiplication Facts of 3	41
	3	Dividing by 3	44
	4	Practice A	47
	5	The Multiplication Table of 4	48
	6	Multiplication Facts of 4	51
	7	Dividing by 4	53
	8	Practice B	55
	9	Practice C	56
		Workbook Pages	60

Chapter		Lesson	Page
Chapter 10 **Money**		Teaching Notes	67
		Chapter Opener	69
	1	Making $1	70
	2	Dollars and Cents	72
	3	Making Change	75
	4	Comparing Money	77
	5	Practice A	79
	6	Adding Money	80
	7	Subtracting Money	83
	8	Practice B	86
		Workbook Pages	88
Chapter 11 **Fractions**		Teaching Notes	97
		Chapter Opener	101
	1	Halves and Fourths	102
	2	Writing Unit Fractions	105
	3	Writing Fractions	107
	4	Fractions that Make 1 Whole	109
	5	Comparing and Ordering Fractions	111
	6	Practice	113
		Review 3	115
		Workbook Pages	117
Chapter 12 **Time**		Teaching Notes	125
		Chapter Opener	127
	1	Telling Time	128
	2	Time Intervals	131
	3	A.M. and P.M.	134
	4	Practice	136
		Workbook Pages	138

Chapter		Lesson	Page
Chapter 13 **Capacity**		Teaching Notes	143
		Chapter Opener	145
	1	Comparing Capacity	146
	2	Units of Capacity	149
	3	Practice	152
		Workbook Pages	153
Chapter 14 **Graphs**		Teaching Notes	157
		Chapter Opener	159
	1	Picture Graphs	160
	2	Bar Graphs	163
	3	Practice	165
		Workbook Pages	167
Chapter 15 **Shapes**		Teaching Notes	171
		Chapter Opener	175
	1	Straight and Curved Sides	176
	2	Polygons	178
	3	Semicircles and Quarter-circles	182
	4	Patterns	184
	5	Solid Shapes	187
	6	Practice	190
		Review 4	192
		Review 5	194
		Workbook Pages	196
Resources		Blackline Masters for 2B	206

Notes

Dimensions Math® Curriculum

The **Dimensions Math®** series is a Pre-Kindergarten to Grade 5 series based on the pedagogy and methodology of math education in Singapore. The main goal of the **Dimensions Math®** series is to help students develop competence and confidence in mathematics.

The series follows the principles outlined in the Singapore Mathematics Framework below.

Pedagogical Approach and Methodology

- Through Concrete-Pictorial-Abstract development, students view the same concepts over time with increasing levels of abstraction.
- Thoughtful sequencing creates a sense of continuity. The content of each grade level builds on that of preceding grade levels. Similarly, lessons build on previous lessons within each grade.
- Group discussion of solution methods encourages expansive thinking.
- Interesting problems and activities provide varied opportunities to explore and apply skills.
- Hands-on tasks and sharing establish a culture of collaboration.
- Extra practice and extension activities encourage students to persevere through challenging problems.
- Variation in pictorial representation (number bonds, bar models, etc.) and concrete representation (straws, linking cubes, base ten blocks, discs, etc.) broaden student understanding.

Each topic is introduced, then thoughtfully developed through the use of a variety of learning experiences, problem solving, student discourse, and opportunities for mastery of skills. This combination of hands-on practice, in-depth exploration of topics, and mathematical variability in teaching methodology allows students to truly master mathematical concepts.

Singapore Mathematics Framework

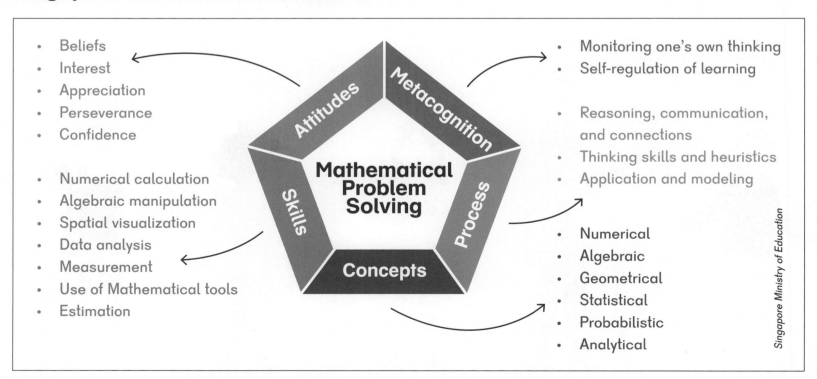

Dimensions Math® Program Materials

Textbooks

Textbooks are designed to help students build a solid foundation in mathematical thinking and efficient problem solving. Careful sequencing of topics, well-chosen problems, and simple graphics foster deep conceptual understanding and confidence. Mental math, problem solving, and correct computation are given balanced attention in all grades. As skills are mastered, students move to increasingly sophisticated concepts within and across grade levels.

Students work through the textbook lessons with the help of five friends: Emma, Alex, Sofia, Dion, and Mei. The characters appear throughout the series and help students develop metacognitive reasoning through questions, hints, and ideas.

A pencil icon ➤ at the end of the textbook lessons links to exercises in the workbooks.

Workbooks

Workbooks provide additional problems that range from basic to challenging. These allow students to independently review and practice the skills they have learned.

Teacher's Guides

Teacher's Guides include lesson plans, mathematical background, games, helpful suggestions, and comprehensive resources for daily lessons.

Tests

Tests contain differentiated assessments to systematically evaluate student progress.

Emma   Alex   Sofia   Dion   Mei

Online Resources

The following can be downloaded from dimensionsmath.com.

- **Blackline Masters** used for various hands-on tasks.

- **Material Lists** for each chapter and lesson, so teachers and classroom helpers can prepare ahead of time.

- **Activities** that can done with students who need more practice or a greater challenge, organized by concept, chapter, and lesson.

- **Standards Alignments** for various states.

Using the Teacher's Guide

This guide is designed to assist in planning daily lessons. It should be considered a helping hand between the curriculum and the classroom. It provides introductory notes on mathematical content, key points, and suggestions for activities. It also includes ideas for differentiation within each lesson, and answers and solutions to textbook and workbook problems.

Each chapter of the guide begins with the following.

- **Overview**

    Includes objectives and suggested number of class periods for each chapter.

- **Notes**

    Highlights key learning points, provides background on math concepts, explains the purpose of certain activities, and helps teachers understand the flow of topics throughout the year.

- **Materials**

    Lists materials, manipulatives, and Blackline Masters used in the Think and Learn sections of the guide. It also includes suggested storybooks. Many common classroom manipulatives are used throughout the curriculum. When a lesson refers to a whiteboard and markers, any writing materials can be used. Blackline Masters can be found at dimensionsmath.com.

The guide goes through the Chapter Openers, Daily Lessons, and Practices of each chapter, and cumulative reviews in the following general format.

- **<u>Chapter Opener</u>**

    Provides talking points for discussion to prepare students for the math concepts to be introduced.

- **<u>Think</u>**

    Offers structure for teachers to guide student inquiry. Provides various methods and activities to solve initial textbook problems or tasks.

- **<u>Learn</u>**

    Guides teachers to analyze student methods from Think to arrive at the main concepts of the lesson through discussion and study of the pictorial representations in the textbook.

- **<u>Do</u>**

    Expands on specific problems with strategies, additional practice, and remediation.

● Activities

Allows students to practice concepts through individual, small group, and whole group hands-on tasks and games, including suggestions for outdoor play (most of which can be modified for a gymnasium or classroom).

Level of difficulty in the games and activities are denoted by the following symbols.

- ● Foundational activities
- ▲ On-level activities
- ★ Challenge or extension activities

● Brain Works

Provides opportunities for students to extend their mathematical thinking.

Discussion is a critical component of each lesson. Teachers are encouraged to let students discuss their reasoning. As each classroom is different, this guide does not anticipate all situations. The following questions can help students articulate their thinking and increase their mastery:

- Why? How do you know?
- Can you explain that?
- Can you draw a picture of that?
- Is your answer reasonable? How do you know?
- How is this task like the one we did before? How is it different?
- What is alike and what is different about…?
- Can you solve that a different way?
- Yes! You're right! How do you know it's true?
- What did you learn before that can help you solve this problem?
- Can you summarize what your classmate shared?
- What conclusion can you draw from the data?

Each lesson is designed to take one day. If your calendar allows, you may choose to spend more than one day on certain lessons. Throughout the guide, there are notes to extend on learning activities to make them more challenging. Lesson structures and activities do not have to conform exactly to what is shown in the guide. Teachers are encouraged to exercise their discretion in using this material in a way that best suits their classes.

Textbooks are designed to last multiple years. Textbook problems with a ▮ (or a blank line for terms) are meant to invite active participation.

Dimensions Math® Scope & Sequence

PKA

Chapter 1
Match, Sort, and Classify

Red and Blue
Yellow and Green
Color Review
Soft and Hard
Rough, Bumpy, and Smooth
Sticky and Grainy
Size — Part 1
Size — Part 2
Sort Into Two Groups
Practice

Chapter 2
Compare Objects

Big and Small
Long and Short
Tall and Short
Heavy and Light
Practice

Chapter 3
Patterns

Movement Patterns
Sound Patterns
Create Patterns
Practice

Chapter 4
Numbers to 5 — Part 1

Count 1 to 5 — Part 1
Count 1 to 5 — Part 2
Count Back

Count On and Back
Count 1 Object
Count 2 Objects
Count Up to 3 Objects
Count Up to 4 Objects
Count Up to 5 Objects
How Many? — Part 1
How Many? — Part 2
How Many Now? — Part 1
How Many Now? — Part 2
Practice

Chapter 5
Numbers to 5 — Part 2

1, 2, 3
1, 2, 3, 4, 5 — Part 1
1, 2, 3, 4, 5 — Part 2
How Many? — Part 1
How Many? — Part 2
How Many Do You See?
How Many Do You See Now?
Practice

Chapter 6
Numbers to 10 — Part 1

0
Count to 10 — Part 1
Count to 10 — Part 2
Count Back
Order Numbers
Count Up to 6 Objects
Count Up to 7 Objects
Count Up to 8 Objects
Count Up to 9 Objects
Count Up to 10 Objects — Part 1

Count Up to 10 Objects — Part 2
How Many?
Practice

Chapter 7
Numbers to 10 — Part 2

6
7
8
9
10
0 to 10
Count and Match — Part 1
Count and Match — Part 2
Practice

PKB

Chapter 8
Ordinal Numbers

First
Second and Third
Fourth and Fifth
Practice

Chapter 9
Shapes and Solids

Cubes, Cylinders, and Spheres
Cubes
Positions
Build with Solids
Rectangles and Circles
Squares
Triangles

Squares, Circles, Rectangles, and Triangles — Part 1
Squares, Circles, Rectangles, and Triangles — Part 2
Practice

Chapter 10
Compare Sets

Match Objects
Which Set Has More?
Which Set Has Fewer?
More or Fewer?
Practice

Chapter 11
Compose and Decompose

Altogether — Part 1
Altogether — Part 2
Show Me
What's the Other Part? — Part 1
What's the Other Part? — Part 2
Practice

Chapter 12
Explore Addition and Subtraction

Add to 5 — Part 1
Add to 5 — Part 2
Two Parts Make a Whole
How Many in All?
Subtract Within 5 — Part 1
Subtract Within 5 — Part 2
How Many Are Left?

Practice

Chapter 13
Cumulative Review

Review 1 Match and Color
Review 2 Big and Small
Review 3 Heavy and Light
Review 4 Count to 5
Review 5 Count 5 Objects
Review 6 0
Review 7 Count Beads
Review 8 Patterns
Review 9 Length
Review 10 How Many?
Review 11 Ordinal Numbers
Review 12 Solids and Shapes
Review 13 Which Set Has More?
Review 14 Which Set Has Fewer?
Review 15 Put Together
Review 16 Subtraction
Looking Ahead 1 Sequencing — Part 1
Looking Ahead 2 Sequencing — Part 2
Looking Ahead 3 Categorizing
Looking Ahead 4 Addition
Looking Ahead 5 Subtraction
Looking Ahead 6 Getting Ready to Write Numerals
Looking Ahead 7 Reading and Math

KA

Chapter 1
Match, Sort, and Classify

Left and Right
Same and Similar
Look for One That Is Different
How Does it Feel?
Match the Things That Go Together
Sort
Practice

Chapter 2
Numbers to 5

Count to 5
Count Things Up to 5
Recognize the Numbers 1 to 3
Recognize the Numbers 4 and 5
Count and Match
Write the Numbers 1 and 2
Write the Number 3
Write the Number 4
Trace and Write 1 to 5
Zero
Picture Graphs
Practice

Chapter 3
Numbers to 10

Count 1 to 10
Count Up to 7 Things
Count Up to 9 Things
Count Up to 10 Things — Part 1

Dimensions Math® Scope & Sequence

Count Up to 10 Things — Part 2
Recognize the Numbers 6 to 10
Write the Numbers 6 and 7
Write the Numbers 8, 9, and 10
Write the Numbers 6 to 10
Count and Write the Numbers 1 to 10
Ordinal Positions
One More Than
Practice

Chapter 4
Shapes and Solids

Curved or Flat
Solid Shapes
Closed Shapes
Rectangles
Squares
Circles and Triangles
Where is It?
Hexagons
Sizes and Shapes
Combine Shapes
Graphs
Practice

Chapter 5
Compare Height, Length, Weight, and Capacity

Comparing Height
Comparing Length
Height and Length — Part 1
Height and Length — Part 2
Weight — Part 1
Weight — Part 2
Weight — Part 3
Capacity — Part 1
Capacity — Part 2
Practice

Chapter 6
Comparing Numbers Within 10

Same and More
More and Fewer
More and Less
Practice — Part 1
Practice — Part 2

KB

Chapter 7
Numbers to 20

Ten and Some More
Count Ten and Some More
Two Ways to Count
Numbers 16 to 20
Number Words 0 to 10
Number Words 11 to 15
Number Words 16 to 20
Number Order
1 More Than or Less Than
Practice — Part 1
Practice — Part 2

Chapter 8
Number Bonds

Putting Numbers Together — Part 1
Putting Numbers Together — Part 2
Parts Making a Whole
Look for a Part
Number Bonds for 2, 3, and 4
Number Bonds for 5
Number Bonds for 6
Number Bonds for 7
Number Bonds for 8
Number Bonds for 9
Number Bonds for 10
Practice — Part 1
Practice — Part 2
Practice — Part 3

Chapter 9
Addition

Introduction to Addition — Part 1
Introduction to Addition — Part 2
Introduction to Addition — Part 3
Addition
Count On — Part 1
Count On — Part 2
Add Up to 3 and 4
Add Up to 5 and 6
Add Up to 7 and 8
Add Up to 9 and 10
Addition Practice
Practice

Chapter 10
Subtraction

Take Away to Subtract — Part 1

Take Away to Subtract — Part 2
Take Away to Subtract — Part 3
Take Apart to Subtract — Part 1
Take Apart to Subtract — Part 2
Count Back
Subtract Within 5
Subtract Within 10 — Part 1
Subtract Within 10 — Part 2
Practice

Chapter 11
Addition and Subtraction

Add and Subtract
Practice Addition and Subtraction
Part-Whole Addition and Subtraction
Add to or Take Away
Put Together or Take Apart
Practice

Chapter 12
Numbers to 100

Count by Tens — Part 1
Count by Tens — Part 2
Numbers to 30
Numbers to 40
Numbers to 50
Numbers to 80
Numbers to 100 — Part 1
Numbers to 100 — Part 2
Count by Fives — Part 1
Count by Fives — Part 2

Practice

Chapter 13
Time

Day and Night
Learning About the Clock
Telling Time to the Hour — Part 1
Telling Time to the Hour — Part 2
Practice

Chapter 14
Money

Coins
Pennies
Nickels
Dimes
Quarters
Practice

1A

Chapter 1
Numbers to 10

Numbers to 10
The Number 0
Order Numbers
Compare Numbers
Practice

Chapter 2
Number Bonds

Make 6
Make 7
Make 8

Make 9
Make 10 — Part 1
Make 10 — Part 2
Practice

Chapter 3
Addition

Addition as Putting Together
Addition as Adding More
Addition with 0
Addition with Number Bonds
Addition by Counting On
Make Addition Stories
Addition Facts
Practice

Chapter 4
Subtraction

Subtraction as Taking Away
Subtraction as Taking Apart
Subtraction by Counting Back
Subtraction with 0
Make Subtraction Stories
Subtraction with Number Bonds
Addition and Subtraction
Make Addition and Subtraction Story Problems
Subtraction Facts
Practice
Review 1

Chapter 5
Numbers to 20

Numbers to 20
Add or Subtract Tens or Ones
Order Numbers to 20

Dimensions Math® Scope & Sequence

Compare Numbers to 20
Addition
Subtraction
Practice

Chapter 6
Addition to 20

Add by Making 10 — Part 1
Add by Making 10 — Part 2
Add by Making 10 — Part 3
Addition Facts to 20
Practice

Chapter 7
Subtraction Within 20

Subtract from 10 — Part 1
Subtract from 10 — Part 2
Subtract the Ones First
Word Problems
Subtraction Facts Within 20
Practice

Chapter 8
Shapes

Solid and Flat Shapes
Grouping Shapes
Making Shapes
Practice

Chapter 9
Ordinal Numbers

Naming Positions
Word Problems
Practice
Review 2

1B

Chapter 10
Length

Comparing Lengths Directly
Comparing Lengths Indirectly
Comparing Lengths with Units
Practice

Chapter 11
Comparing

Subtraction as Comparison
Making Comparison
   Subtraction Stories
Picture Graphs
Practice

Chapter 12
Numbers to 40

Numbers to 40
Tens and Ones
Counting by Tens and Ones
Comparing
Practice

Chapter 13
Addition and Subtraction Within 40

Add Ones
Subtract Ones
Make the Next Ten
Use Addition Facts
Subtract from Tens
Use Subtraction Facts
Add Three Numbers
Practice

Chapter 14
Grouping and Sharing

Adding Equal Groups
Sharing
Grouping
Practice

Chapter 15
Fractions

Halves
Fourths
Practice
Review 3

Chapter 16
Numbers to 100

Numbers to 100
Tens and Ones
Count by Ones or Tens
Compare Numbers to 100
Practice

Chapter 17
Addition and Subtraction Within 100

Add Ones — Part 1
Add Tens
Add Ones — Part 2
Add Tens and Ones — Part 1
Add Tens and Ones — Part 2
Subtract Ones — Part 1
Subtract from Tens
Subtract Ones — Part 2
Subtract Tens

Subtract Tens and Ones —
　　Part 1
Subtract Tens and Ones —
　　Part 2
Practice

Chapter 18
Time

Telling Time to the Hour
Telling Time to the Half Hour
Telling Time to the 5 Minutes
Practice

Chapter 19
Money

Coins
Counting Money
Bills
Shopping
Practice
Review 4

2A

Chapter 1
Numbers to 1,000

Tens and Ones
Counting by Tens or Ones
Comparing Tens and Ones
Hundreds, Tens, and Ones
Place Value
Comparing Hundreds, Tens,
　　and Ones
Counting by Hundreds, Tens,
　　or Ones
Practice

Chapter 2
Addition and Subtraction — Part 1

Strategies for Addition
Strategies for Subtraction
Parts and Whole
Comparison
Practice

Chapter 3
Addition and Subtraction — Part 2

Addition Without Regrouping
Subtraction Without
　　Regrouping
Addition with Regrouping
　　Ones
Addition with Regrouping
　　Tens
Addition with Regrouping
　　Tens and Ones
Practice A
Subtraction with Regrouping
　　from Tens
Subtraction with Regrouping
　　from Hundreds
Subtraction with Regrouping
　　from Two Places
Subtraction with Regrouping
　　across Zeros
Practice B
Practice C

Chapter 4
Length

Centimeters
Estimating Length in
　　Centimeters
Meters
Estimating Length in Meters
Inches
Using Rulers
Feet
Practice

Chapter 5
Weight

Grams
Kilograms
Pounds
Practice
Review 1

Chapter 6
Multiplication and Division

Multiplication — Part 1
Multiplication — Part 2
Practice A
Division — Part 1
Division — Part 2
Multiplication and Division
Practice B

Chapter 7
Multiplication and Division of 2, 5, and 10

The Multiplication Table of 5
Multiplication Facts of 5
Practice A
The Multiplication Table of 2
Multiplication Facts of 2
Practice B
The Multiplication Table of 10
Dividing by 2

Dimensions Math® Scope & Sequence

Dividing by 5 and 10
Practice C
Word Problems
Review 2

2B

Chapter 8
Mental Calculation

Adding Ones Mentally
Adding Tens Mentally
Making 100
Adding 97, 98, or 99
Practice A
Subtracting Ones Mentally
Subtracting Tens Mentally
Subtracting 97, 98, or 99
Practice B
Practice C

Chapter 9
Multiplication and Division of 3 and 4

The Multiplication Table of 3
Multiplication Facts of 3
Dividing by 3
Practice A
The Multiplication Table of 4
Multiplication Facts of 4
Dividing by 4
Practice B
Practice C

Chapter 10
Money

Making $1
Dollars and Cents
Making Change
Comparing Money
Practice A
Adding Money
Subtracting Money
Practice B

Chapter 11
Fractions

Halves and Fourths
Writing Unit Fractions
Writing Fractions
Fractions that Make 1 Whole
Comparing and Ordering Fractions
Practice
Review 3

Chapter 12
Time

Telling Time
Time Intervals
A.M. and P.M.
Practice

Chapter 13
Capacity

Comparing Capacity
Units of Capacity
Practice

Chapter 14
Graphs

Picture Graphs
Bar Graphs
Practice

Chapter 15
Shapes

Straight and Curved Sides
Polygons
Semicircles and Quarter-circles
Patterns
Solid Shapes
Practice
Review 4
Review 5

3A

Chapter 1
Numbers to 10,000

Numbers to 10,000
Place Value — Part 1
Place Value — Part 2
Comparing Numbers
The Number Line
Practice A
Number Patterns
Rounding to the Nearest Thousand
Rounding to the Nearest Hundred
Rounding to the Nearest Ten
Practice B

Chapter 2
Addition and Subtraction — Part 1

Mental Addition — Part 1
Mental Addition — Part 2
Mental Subtraction — Part 1
Mental Subtraction — Part 2
Making 100 and 1,000
Strategies for Numbers Close to Hundreds
Practice A
Sum and Difference
Word Problems — Part 1
Word Problems — Part 2
2-Step Word Problems
Practice B

Chapter 3
Addition and Subtraction — Part 2

Addition with Regrouping
Subtraction with Regrouping — Part 1
Subtraction with Regrouping — Part 2
Estimating Sums and Differences — Part 1
Estimating Sums and Differences — Part 2
Word Problems
Practice

Chapter 4
Multiplication and Division

Looking Back at Multiplication
Strategies for Finding the Product
Looking Back at Division
Multiplying and Dividing with 0 and 1
Division with Remainders
Odd and Even Numbers
Word Problems — Part 1
Word Problems — Part 2
2-Step Word Problems
Practice
Review 1

Chapter 5
Multiplication

Multiplying Ones, Tens, and Hundreds
Multiplication Without Regrouping
Multiplication with Regrouping Tens
Multiplication with Regrouping Ones
Multiplication with Regrouping Ones and Tens
Practice A
Multiplying a 3-Digit Number with Regrouping Once
Multiplication with Regrouping More Than Once
Practice B

Chapter 6
Division

Dividing Tens and Hundreds
Dividing a 2-Digit Number by 2 — Part 1
Dividing a 2-Digit Number by 2 — Part 2
Dividing a 2-Digit Number by 3, 4, and 5
Practice A
Dividing a 3-Digit Number by 2
Dividing a 3-Digit Number by 3, 4, and 5
Dividing a 3-Digit Number, Quotient is 2 Digits
Practice B

Chapter 7
Graphs and Tables

Picture Graphs and Bar Graphs
Bar Graphs and Tables
Practice
Review 2

3B

Chapter 8
Multiplying and Dividing with 6, 7, 8, and 9

The Multiplication Table of 6
The Multiplication Table of 7
Multiplying by 6 and 7
Dividing by 6 and 7
Practice A
The Multiplication Table of 8

Dimensions Math® Scope & Sequence

The Multiplication Table of 9
Multiplying by 8 and 9
Dividing by 8 and 9
Practice B

Chapter 9
Fractions — Part 1

Fractions of a Whole
Fractions on a Number Line
Comparing Fractions with Like Denominators
Comparing Fractions with Like Numerators
Practice

Chapter 10
Fractions — Part 2

Equivalent Fractions
Finding Equivalent Fractions
Simplifying Fractions
Comparing Fractions — Part 1
Comparing Fractions — Part 2
Practice A
Adding and Subtracting Fractions — Part 1
Adding and Subtracting Fractions — Part 2
Practice B

Chapter 11
Measurement

Meters and Centimeters
Subtracting from Meters
Kilometers
Subtracting from Kilometers
Liters and Milliliters
Kilograms and Grams

Word Problems
Practice
Review 3

Chapter 12
Geometry

Circles
Angles
Right Angles
Triangles
Properties of Triangles
Properties of Quadrilaterals
Using a Compass
Practice

Chapter 13
Area and Perimeter

Area
Units of Area
Area of Rectangles
Area of Composite Figures
Practice A
Perimeter
Perimeter of Rectangles
Area and Perimeter
Practice B

Chapter 14
Time

Units of Time
Calculating Time — Part 1
Practice A
Calculating Time — Part 2
Calculating Time — Part 3
Calculating Time — Part 4
Practice B

Chapter 15
Money

Dollars and Cents
Making $10
Adding Money
Subtracting Money
Word Problems
Practice
Review 4
Review 5

4A

Chapter 1
Numbers to One Million

Numbers to 100,000
Numbers to 1,000,000
Number Patterns
Comparing and Ordering Numbers
Rounding 5-Digit Numbers
Rounding 6-Digit Numbers
Calculations and Place Value
Practice

Chapter 2
Addition and Subtraction

Addition
Subtraction
Other Ways to Add and Subtract — Part 1
Other Ways to Add and Subtract — Part 2
Word Problems

Practice

Chapter 3
Multiples and Factors

Multiples
Common Multiples
Factors
Prime Numbers and
 Composite Numbers
Common Factors
Practice

Chapter 4
Multiplication

Mental Math for Multiplication
Multiplying by a 1-Digit
 Number — Part 1
Multiplying by a 1-Digit
 Number — Part 2
Practice A
Multiplying by a Multiple of 10
Multiplying by a 2-Digit
 Number — Part 1
Multiplying by a 2-Digit
 Number — Part 2
Practice B

Chapter 5
Division

Mental Math for Division
Estimation and Division
Dividing 4-Digit Numbers
Practice A
Word Problems
Challenging Word Problems
Practice B
Review 1

Chapter 6
Fractions

Equivalent Fractions
Comparing and Ordering
 Fractions
Improper Fractions and Mixed
 Numbers
Practice A
Expressing an Improper
 Fraction as a Mixed
 Number
Expressing a Mixed Number
 as an Improper Fraction
Fractions and Division
Practice B

Chapter 7
Adding and Subtracting Fractions

Adding and Subtracting
 Fractions — Part 1
Adding and Subtracting
 Fractions — Part 2
Adding a Mixed Number and
 a Fraction
Adding Mixed Numbers
Subtracting a Fraction from
 a Mixed Number
Subtracting Mixed Numbers
Practice

Chapter 8
Multiplying a Fraction and a Whole Number

Multiplying a Unit Fraction
 by a Whole Number
Multiplying a Fraction by a
 Whole Number — Part 1
Multiplying a Fraction by a
 Whole Number — Part 2
Fraction of a Set
Multiplying a Whole Number
 by a Fraction — Part 1
Multiplying a Whole Number
 by a Fraction — Part 2
Word Problems — Part 1
Word Problems — Part 2
Practice

Chapter 9
Line Graphs and Line Plots

Line Graphs
Drawing Line Graphs
Line Plots
Practice
Review 2

4B

Chapter 10
Measurement

Metric Units of Measurement
Customary Units of Length
Customary Units of Weight
Customary Units of Capacity
Units of Time
Practice A
Fractions and Measurement
 — Part 1
Fractions and Measurement
 — Part 2
Practice B

Dimensions Math® Scope & Sequence

Chapter 11
Area and Perimeter

Area of Rectangles — Part 1
Area of Rectangles — Part 2
Area of Composite Figures
Perimeter — Part 1
Perimeter — Part 2
Practice

Chapter 12
Decimals

Tenths — Part 1
Tenths — Part 2
Hundredths — Part 1
Hundredths — Part 2
Expressing Decimals as Fractions in Simplest Form
Expressing Fractions as Decimals
Practice A
Comparing and Ordering Decimals
Rounding Decimals
Practice B

Chapter 13
Addition and Subtraction of Decimals

Adding and Subtracting Tenths
Adding Tenths with Regrouping
Subtracting Tenths with Regrouping
Practice A
Adding Hundredths
Subtracting from 1 and 0.1
Subtracting Hundredths
Money, Decimals, and Fractions

Practice B
Review 3

Chapter 14
Multiplication and Division of Decimals

Multiplying Tenths and Hundredths
Multiplying Decimals by a Whole Number — Part 1
Multiplying Decimals by a Whole Number — Part 2
Practice A
Dividing Tenths and Hundredths
Dividing Decimals by a Whole Number — Part 1
Dividing Decimals by a Whole Number — Part 2
Dividing Decimals by a Whole Number — Part 3
Practice B

Chapter 15
Angles

The Size of Angles
Measuring Angles
Drawing Angles
Adding and Subtracting Angles
Reflex Angles
Practice

Chapter 16
Lines and Shapes

Perpendicular Lines
Parallel Lines
Drawing Perpendicular and Parallel Lines
Quadrilaterals

Lines of Symmetry
Symmetrical Figures and Patterns
Practice

Chapter 17
Properties of Cuboids

Cuboids
Nets of Cuboids
Faces and Edges of Cuboids
Practice
Review 4
Review 5

5A

Chapter 1
Whole Numbers

Numbers to One Billion
Multiplying by 10, 100, and 1,000
Dividing by 10, 100, and 1,000
Multiplying by Tens, Hundreds, and Thousands
Dividing by Tens, Hundreds, and Thousands
Practice

Chapter 2
Writing and Evaluating Expressions

Expressions with Parentheses
Order of Operations — Part 1
Order of Operations — Part 2

Other Ways to Write and
    Evaluate Expressions
Word Problems — Part 1
Word Problems — Part 2
Practice

Chapter 3
Multiplication and Division

Multiplying by a 2-digit
    Number — Part 1
Multiplying by a 2-digit
    Number — Part 2
Practice A
Dividing by a Multiple of Ten
Divide a 2-digit Number by a
    2-digit Number
Divide a 3-digit Number by a
    2-digit Number — Part 1
Divide a 3-digit Number by a
    2-digit Number — Part 2
Divide a 4-digit Number by a
    2-digit Number
Practice B

Chapter 4
Addition and Subtraction of Fractions

Fractions and Division
Adding Unlike Fractions
Subtracting Unlike Fractions
Practice A
Adding Mixed Numbers
    — Part 1
Adding Mixed Numbers
    — Part 2
Subtracting Mixed Numbers
    — Part 1

Subtracting Mixed Numbers
    — Part 2
Practice B
Review 1

Chapter 5
Multiplication of Fractions

Multiplying a Fraction by a
    Whole Number
Multiplying a Whole Number
    by a Fraction
Word Problems — Part 1
Practice A
Multiplying a Fraction by a
    Unit Fraction
Multiplying a Fraction by a
    Fraction — Part 1
Multiplying a Fraction by a
    Fraction — Part 2
Multiplying Mixed Numbers
Word Problems — Part 2
Fractions and Reciprocals
Practice B

Chapter 6
Division of Fractions

Dividing a Unit Fraction by a
    Whole Number
Dividing a Fraction by a
    Whole Number
Practice A
Dividing a Whole Number by
    a Unit Fraction
Dividing a Whole Number by
    a Fraction
Word Problems
Practice B

Chapter 7
Measurement

Fractions and Measurement
    Conversions
Fractions and Area
Practice A
Area of a Triangle — Part 1
Area of a Triangle — Part 2
Area of Complex Figures
Practice B

Chapter 8
Volume of Solid Figures

Cubic Units
Volume of Cuboids
Finding the Length of an Edge
Practice A
Volume of Complex Shapes
Volume and Capacity — Part 1
Volume and Capacity — Part 2
Practice B
Review 2

5B

Chapter 9
Decimals

Thousandths
Place Value to Thousandths
Comparing Decimals
Rounding Decimals
Practice A
Multiply Decimals by 10, 100,
    and 1,000
Divide Decimals by 10, 100,
    and 1,000

Dimensions Math® Scope & Sequence

Conversion of Measures
Mental Calculation
Practice B

Chapter 10
The Four Operations of Decimals

Adding Decimals to Thousandths
Subtracting Decimals
Multiplying by 0.1 or 0.01
Multiplying by a Decimal
Practice A
Dividing by a Whole Number — Part 1
Dividing by a Whole Number — Part 2
Dividing a Whole Number by 0.1 and 0.01
Dividing a Whole Number by a Decimal
Practice B

Chapter 11
Geometry

Measuring Angles
Angles and Lines
Classifying Triangles
The Sum of the Angles in a Triangle
The Exterior Angle of a Triangle
Classifying Quadrilaterals
Angles of Quadrilaterals — Part 1
Angles of Quadrilaterals — Part 2

Drawing Triangles and Quadrilaterals
Practice

Chapter 12
Data Analysis and Graphs

Average — Part 1
Average — Part 2
Line Plots
Coordinate Graphs
Straight Line Graphs
Practice
Review 3

Chapter 13
Ratio

Finding the Ratio
Equivalent Ratios
Finding a Quantity
Comparing Three Quantities
Word Problems
Practice

Chapter 14
Rate

Finding the Rate
Rate Problems — Part 1
Rate Problems — Part 2
Word Problems
Practice

Chapter 15
Percentage

Meaning of Percentage
Expressing Percentages as Fractions

Percentages and Decimals
Expressing Fractions as Percentages
Practice A
Percentage of a Quantity
Word Problems
Practice B
Review 4
Review 5

Chapter 8 Mental Calculation

Overview

Suggested number of class periods: 10–11

Lesson		Page	Resources	Objectives
	Chapter Opener	p. 5	TB: p. 1	Investigate mental math strategies.
1	Adding Ones Mentally	p. 6	TB: p. 2 WB: p. 1	Add ones to a two-digit or three-digit number mentally.
2	Adding Tens Mentally	p. 8	TB: p. 4 WB: p. 3	Add a two-digit multiple of 10 to a two-digit or three-digit number mentally.
3	Making 100	p. 10	TB: p. 6 WB: p. 5	Find the complement of a number that makes 100.
4	Adding 97, 98, or 99	p. 12	TB: p. 8 WB: p. 7	Add 97, 98, or 99 to a two-digit or three-digit number mentally.
5	Practice A	p. 14	TB: p. 10 WB: p. 9	Practice addition using mental math strategies.
6	Subtracting Ones Mentally	p. 16	TB: p. 12 WB: p. 13	Subtract ones from a two-digit or three-digit number mentally.
7	Subtracting Tens Mentally	p. 18	TB: p. 14 WB: p. 15	Subtract a multiple of 10 from a two-digit or three-digit number mentally.
8	Subtracting 97, 98, or 99	p. 20	TB: p. 16 WB: p. 17	Subtract 97, 98, or 99 from a two-digit or three-digit number mentally.
9	Practice B	p. 22	TB: p. 18 WB: p. 19	Practice subtraction using mental math strategies.
10	Practice C	p. 24	TB: p. 20 WB: p. 23	Practice addition and subtraction using mental math.
	Workbook Solutions	p. 26		

Chapter 8 Mental Calculation

In **Dimensions Math® 2A** Chapter 1, students added and subtracted 1, 2, 3, 10, 20, and 30 with regrouping.

In **Dimensions Math® 1B** Chapter 14, students learned the following mental math strategies for adding and subtracting one-digit and two-digit numbers:

Add by making the next 10	$27 + 8 = 35$	$27 + 8 = 30 + 5 = 35$ with 8 split into 3 and 5	
Add using known facts to 20	Students know from memory that $7 + 8 = 15$	$27 + 8 = 20 + 15 = 35$ with 27 split into 20 and 7	
Subtract from a 10	$35 - 8$	$35 - 8 = 25 + 2 = 27$ with 35 split into 25 and 10; $10 - 8 = 2$	$35 - 8 = 5 + 22 = 27$ with 35 split into 5 and 30; $30 - 8 = 22$
Subtract by using known facts to 20	Students know from memory that $15 - 8 = 7$ or that $8 + \_\_\_ = 15$	$35 - 8 = 20 + 7 = 27$ with 35 split into 20 and 15; $15 - 8 = 7$	

In **Dimensions Math® 1B** Chapter 17, students learned the following mental math strategies for adding and subtracting two-digit numbers:

Add a two-digit number and a multiple of ten by adding the tens	$62 + 30 = 92$	$62 + 30 = 92$ with 62 split into 60 and 2
Add 2 two-digit numbers by adding tens and ones	$45 + 26 = 71$	$45 + 26 = 45 + 20 + 6 = 71$ with 26 split into 20 and 6
Subtract a multiple of ten from a two-digit number	$62 - 30 = 32$	$62 - 30 = 60 - 30 + 2 = 32$ with 62 split into 60 and 2
Subtract a two-digit number	$65 - 28 = 37$	$65 - 28 = 65 - 20 - 8 = 37$ with 28 split into 20 and 8

Chapter 8 Mental Calculation

In this chapter, these strategies will be reinforced and extended to three-digit numbers.

For example, students may use strategies from **Dimensions Math® 1B** with a three-digit number:

762 + 9 = 761 + 10 = 771
(761) (1)

762 + 9 = 770 + 1 = 771
(8) (1)

762 + 9 = 760 + 11 = 771
(760) (2)

Additionally, students will learn to add or subtract a number close to 100:

Add 97, 98, or 99 by making the next ten	99 + 52 = 151	99 + 52 = 100 + 51 = 151 (1) (51)
Add 97, 98, or 99 by adding 100 and subtracting 1, 2, or 3	99 + 52 = 151	100 (99) (1) 99 + 52 = 100 + 52 − 1 = 151
Subtract 97, 98, or 99 from 100	480 − 99 = 381	480 − 99 = 380 + 1 = 381 (380) (100) − 99 = 1
Subtract 97, 98, or 99 by subtracting 100 and adding 1, 2, or 3	480 − 99 = 381	100 (99) (1) 480 − 99 = 480 − 100 + 1 = 381

Mental math refers to mental strategies that leverage number sense. It is a way to make difficult computation easier. It does not mean that a student isn't allowed to write down numbers while computing. However, as students master the strategies, it is not necessary to have them include an explicit written decomposition for every problem.

Although students will model the strategies shown in the textbook initially with place-value discs, they should transition quickly to visualizing the number bonds and be able to find the answers mentally when practicing.

Additionally, students should be encouraged to use mental strategies throughout their math education when possible. Teachers should look for opportunities to point them out in future lessons.

Chapter 8 Mental Calculation

Materials

- 10-sided dice
- Counters or linking cubes
- Die with modified sides labeled: 97, 97, 98, 98, 99, 99
- Dry erase markers
- Dry erase sleeves
- Index cards
- Recording sheet
- Place-value discs
- Playing cards
- Whiteboards

Blackline Masters

- Double Hundred Chart
- Hundred Chart
- Number Cards

Storybooks

Each of these Greg Tang books have riddles that help readers find easier ways to count, group, add, and subtract:

- *Math Appeal*
- *Math Potatoes*
- *The Grapes of Math*

And just for fun:

- *Math Curse* by Jon Scieszka

Chapter Opener

Objective

- Investigate mental math strategies.

Have students recall strategies from grade 1 that might help them add the prices of items without using the algorithm.

Ask questions like:

- What are the two easiest prices to add mentally? Why?
- Is there a way to organize the numbers to make them easier to add?
- Is there a way to add 100 easily and then see how adding 99 is almost the same?

Lesson 1 Adding Ones Mentally

Objective

- Add ones to a two-digit or three-digit number mentally.

Lesson Materials

- Place-value discs

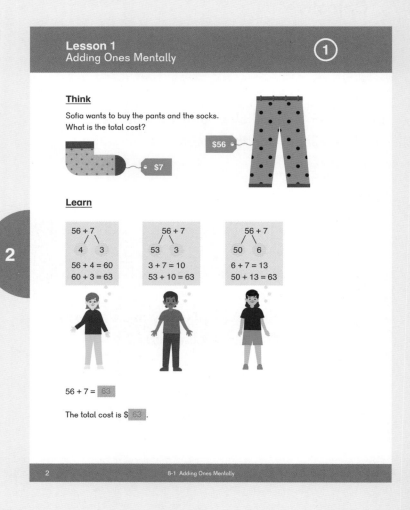

Think

Pose the **Think** problem and provide students with place-value discs.

Have students think of a strategy to solve the problem and show it with place-value discs or number bonds while they share their solutions.

Learn

Model the three strategies used in the textbook with place-value discs. Discuss the strategies with students.

Ask:

- Do you remember using these strategies in grade 1?
- How are Emma and Alex's strategies similar?
- How can we show the strategies with the discs?

Remind students when we work with discs and have 10 ones, we can regroup them into a ten.

Discuss with students why Emma, Alex, and Mei's strategies can be easier to do mentally than working with the discs and regrouping.

Emma and Alex split an addend to make a ten or the next ten.

Mei knows that 6 + 7 is 13. She can easily add $13 and $50.

6   Teacher's Guide 2B Chapter 8   © 2017 Singapore Math Inc.

Do

2 Have students discuss the different strategies with three-digit numbers. Students should not need the discs to solve the problems.

(b) Draw attention to how Sofia rearranged the numbers.

3 Have students share their strategies for solving some of the problems.

Activity

▲ 501 Up

Materials: Recording sheet for each player, Number Cards (BLM) 0 to 9 and 10, 20, 30, 40, 50, 60, 70, 80, 90

Shuffle the Number Cards (BLM) and place them facedown in a pile. Players each begin with 50 points on their recording sheet.

On each turn, a player draws a card and adds that number to his current total.

The first player to go above 501 is the winner.

Exercise 1 • page 1

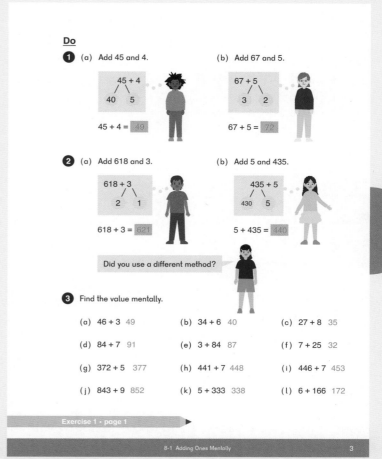

Lesson 2 Adding Tens Mentally

Objective

- Add a two-digit multiple of 10 to a two-digit or three-digit number mentally.

Lesson Materials

- Place-value discs

Think

Pose the **Think** problem. Ask students what they notice about the cost of the jacket and the shoes (there are no ones, they are both tens).

Have students think of a strategy to solve the problem and show it with place-value discs or number bonds while they share their solutions.

Learn

Model the four strategies used in the textbook with place-value discs. Discuss the strategies with students.

Ask:

- How are these strategies different from the ones we used in the last lesson? (They are adding numbers with hundreds and tens, but no ones.)
- How are Alex and Emma's strategies similar?
- Why does Sofia's strategy work?

Alex and Emma split an addend to make a hundred or the next hundred.

Dion knows that 70 + 80 is 150. He can easily add $100 and $150.

Sofia is thinking about renaming hundreds as tens.

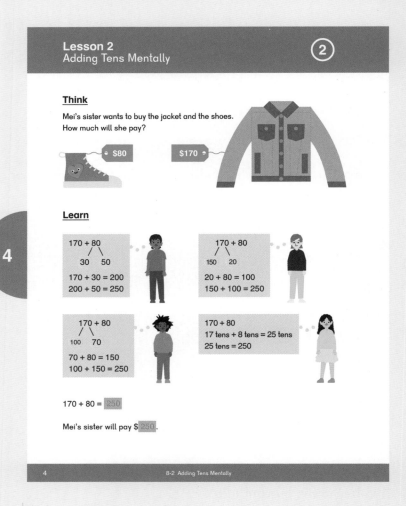

Do

Have students share their different strategies. Write them on the board so they can discuss each other's strategies.

❸ Have students share some of the strategies they used to find the value.

(f) and (i) Remind students that they can rearrange the numbers if it makes it easier for them to solve the problems.

Activity

▲ **Choral Counting**

Using your thumb to point up or down, have students chorally count on and back.

Example:

- "Let's count by fives starting at 195, first number?" Class: "195."
- Point thumb up. Class responds, "200."
- Point thumb up again. Class responds, "205."
- Point thumb down. Class responds, "200."

Or:

- "Let's count by threes starting at 42. First number?" Class: "42."
- Point thumb up. Class responds, "45."
- Point thumb up. Class responds, "48."
- Point thumb up. Class responds, "51."

Exercise 2 • page 3

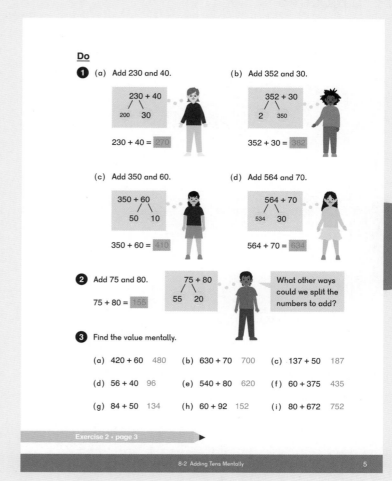

Lesson 3 Making 100

Objective
- Find the complement of a number that makes 100.

Lesson Materials
- Number Cards (BLM) as shown in **Think**, or index cards with those numbers written on them
- Hundred Chart (BLM) in a dry erase sleeve for each student

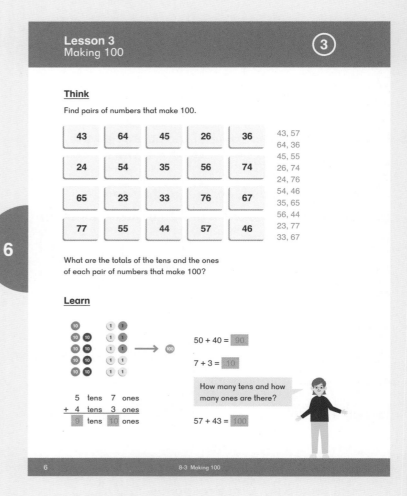

Think

Provide students with a set of Number Cards (BLM) with numbers from the **Think** activity.

Have them find the pairs that sum to 100 and discuss patterns that they see.

Ask students, "What do you notice about the sum of the ones and the sum of the tens in each case?"

Provide each student with a Hundred Chart (BLM) in a dry erase sleeve. Have them choose a card from the set and circle the number on the Hundred Chart (BLM). Have them shade the number that will make 100 with the circled number.

Encourage students to shade the number of ones needed to the next ten, then the number of tens needed to make 100.

Learn

Discuss Emma's question.

Students should note that the sum is always 90 when the tens are added together. The sum of the ones is always 10.

Each pair that sums to 100 can be thought of as 90 + 10.

Do

❶ – ❷ Have students solve by asking themselves, "How many more tens will make 90?" and, "How many more ones will make 10?"

❸ Suggest that students might think of these problems as addition equations adding up to 100, such as, "84 and what number make 100?" They can also use the strategy suggested for ❶ – ❷.

❹ Note that the word "more" in the problem does not necessarily mean "add." If students struggle with the problem, draw a bar model to show this is a missing part or subtraction situation.

Activity

▲ **Memory**

Materials: Cards from **Think** and additional Number Cards (BLM) that sum to 100

Lay Number Cards (BLM) facedown in a 5 by 4 array. Players take turns turning over two cards. If the cards sum to 100, players have found a match and keep the cards.

If the two cards do not add up to 100, the cards are turned facedown again and the player's turn is over.

The player with the most cards at the end of the game is the winner.

● Modify the game for students who are struggling by keeping the cards faceup and having them take turns to find pairs.

Exercise 3 • page 5

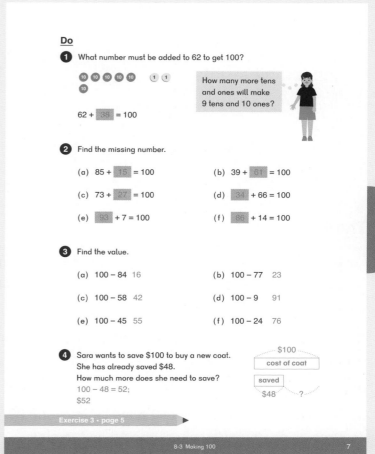

Lesson 4 Adding 97, 98, or 99

Objective

- Add 97, 98, or 99 to a two-digit or three-digit number mentally.

Lesson Materials

- Double Hundred Chart (BLM) in a dry erase sleeve for each student
- Place-value discs

Think

Pose the **Think** problem of how much Mei's items cost. Give students time to think about and solve the problem on their own, and then share their strategies.

Ask students, "What do you notice about the numbers in the problem?"

Provide each student with a Double Hundred Chart (BLM) in a dry erase sleeve and have them shade 99 squares on one hundred chart and 46 on the other. They should see that if they move just one from the 46, then the 99 is 100. Moving one decreases the 46 to 45.

Learn

Discuss Alex's and Emma's strategies. Alex is making the next ten or hundred.

Ask students:

- Why did Emma add 100?
- Why is she subtracting 1? (Students sometimes call Emma's the "over-adding" strategy.)

Use a Double Hundred Chart (BLM) to help students visualize Emma's strategy.

Give students some three digit-numbers and place-value discs to represent these numbers. Then have them add 97, 98, or 99 to these numbers by adding one 100-disc and removing the correct amount of 1-discs. For example, to solve 637 + 98, add one 100-disc and remove two 1-discs to arrive at discs representing 735.

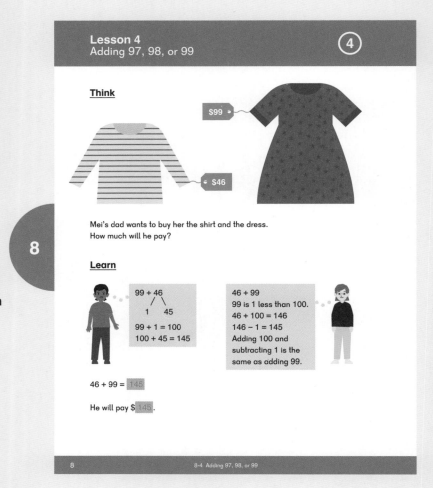

★ Ask students, "Can you do the same thing when adding 9?" (+ 10 − 1) "Adding 29?" (+ 30 − 1)

Do

① – ② These problems encourage students to use a specific strategy.

③ Allow students to use the strategy that works best for them. Have them share some of the strategies they used.

Activity

▲ **Add 97, 98, or 99**

Materials: Three 10-sided dice or decks of Number Cards (BLM) 0 to 9, die with modified sides: 97, 97, 98, 98, 99, 99

Each player tries to make the greatest sum on a given round. Players take turn rolling the dice. On each roll, they make a three-digit number using the numbers from the three 10-sided dice and add the value from the modified 97–98–99 die. The player with the greatest total gets a point. Play 5 rounds.

Then, players play 5 more rounds, trying to make the least sum. The player with the most points after 10 rounds is the winner.

★ Extend by using one 10-sided die as the hundreds place and the modified die as the tens and ones. Use the remaining two 10-sided dice to make a two-digit number and add.

Example:

299 + 59 =

Exercise 4 • page 7

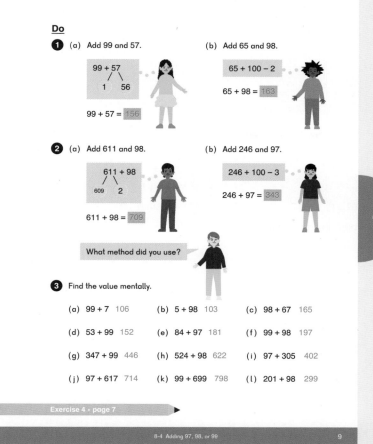

Lesson 5 Practice A

Objective
- Practice addition using mental math strategies.

After students complete the **Practice** in the textbook, have them continue to practice mental addition with activities from the chapter.

Ensure that students are using mental math strategies and allow them to use the strategy that works best for them.

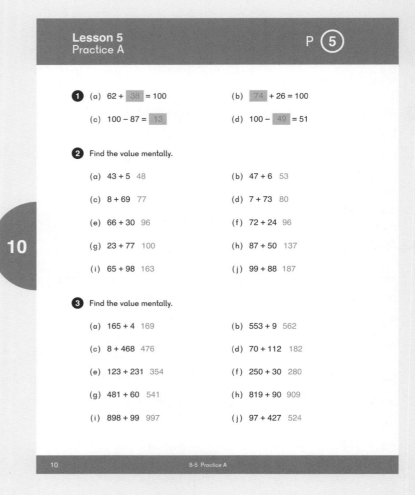

4 – **7** Students can draw bar models as needed.

Students have at least 3 mental math strategies they might use:

- 100 + 47 = 147
  147 + 99 = 147 + 100 − 1
- 100 + 100 + 46
- 100 + 100 + 47 − 1

Activity

▲ Three in a Row – to 100!

Materials: Hundred Chart (BLM) in dry erase sleeve, multiple sets of Number Cards (BLM) or playing cards 0 to 9, dry erase markers

Shuffle the cards and deal 4 cards to each player. Players take turns arranging their numbers into 2 two-digit numbers that they then add together.

If a player is dealt the four cards shown above, she can do 27 + 19, which is 46, so she marks 46 on the Hundred Chart (BLM).

Play continues until a player has three marks in a row, column, or diagonal.

If it is not possible to make two numbers with a sum less than 100, the player may use only 3 of the cards.

If a number is already marked, the digits have to be rearranged to make a different sum.

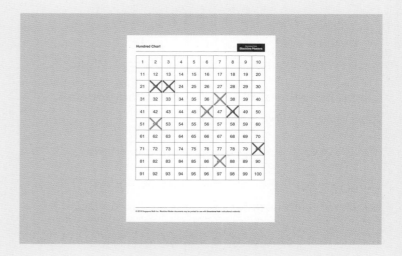

Exercise 5 • page 9

Lesson 6 Subtracting Ones Mentally

Objective

- Subtract ones from a two-digit or three-digit number mentally.

Lesson Materials

- Place-value discs
- Counters or linking cubes

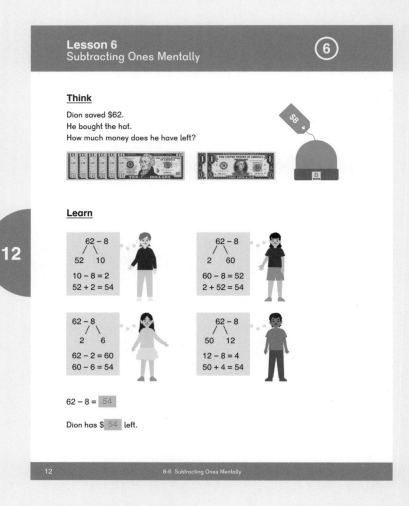

Think

Pose the **Think** problem of how much money Dion has left after he buys the hat.

Have students think of a strategy to solve the problem and show it with place-value discs or number bonds while they share their solutions.

Learn

Model the four strategies used in the textbook with number bonds. Discuss the strategies with students.

Ask:

- Do you remember using these strategies in grade 1?
- How are Emma and Mei's strategies similar?

Emma uses the strategy of subtracting from ten, relying on number bonds to ten.

Mei's strategy is to subtract from the 6 tens.

Sofia breaks the subtrahend into two parts. This can be shown with counters or with linking cubes.

Alex is more comfortable using facts. He knows he can subtract 8 from 12.

Do

❶—❷ Have students discuss the different strategies.

❸ Have students share their different strategies. Write them on the board so they can discuss each other's strategies. Encourage students who finished faster to solve the same problems using a different strategy.

Exercise 6 • page 13

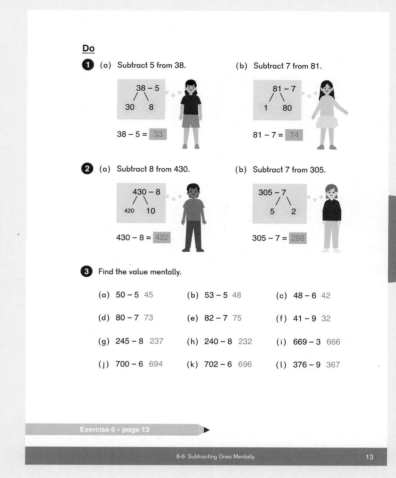

Lesson 7 Subtracting Tens Mentally

Objective

- Subtract a multiple of 10 from a two-digit or three-digit number mentally.

Lesson Materials

- Place-value discs

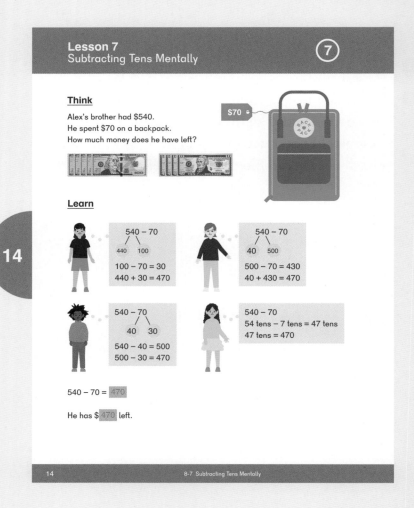

Think

Pose the **Think** problem. Ask students what they notice about the amount of money Alex's brother has and the cost of the backpack.

Possible student responses:

- There are no ones.
- They both only have tens.

Have students work the problem with place-value discs or number bonds and share their solutions.

Learn

Discuss the four strategies used in the textbook.

Ask:

- How are these strategies different from the ones we used in the previous lesson?
- How are Mei and Emma's strategies similar?
- Why does Sofia's strategy work?

Mei is using a similar strategy from the previous lesson, but now with a three-digit number. Instead of subtracting from the ten, she subtracts from the hundred.

Emma and Dion are also using similar strategies from the previous lesson, now applied to three-digit numbers.

For Sofia's strategy, note that whether the problem is 54 oranges − 7 oranges, 54 shoes − 7 shoes, or 54 tens − 7 tens, the answer is 47 of that object.

Do

2 Have students share their different strategies. Write them on the board so other students can understand and discuss another student's strategy.

Activity

▲ 501 Down

Materials: Recording sheet for each player, Number Cards (BLM) 0 to 9 and 10, 20, 30, 40, 50, 60, 70, 80, 90

Shuffle the Number Cards (BLM) and place them facedown in a pile. Players each begin with 501 points on their recording sheets.

On each turn, a player draws a card and subtracts that number from her current total.

The player whose number to be subtracted is greater than her current total is the winner. (For example, "I have 5 and I need to subtract 7. I don't have enough to subtract so I win.")

Exercise 7 • page 15

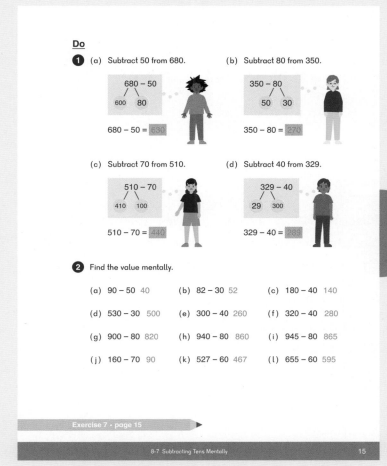

Lesson 8 Subtracting 97, 98, or 99

Objective

- Subtract 97, 98, or 99 from a two-digit or three-digit number mentally.

Lesson Materials

- Place-value discs

Think

Pose the **Think** problem. Provide students with place-value discs and adequate time to work on the problem. Have them share their strategies.

Ask students:

- What do you notice about the numbers in the problem?
- What strategies did we learn for adding 99 that could help us with subtracting 99?

Learn

Discuss Sofia and Dion's strategies. Sofia is subtracting 99 from 100.

Ask students:

- Why did Dion subtract 100 instead of 99?
- Why is he adding 1? (Students sometimes call Dion's the "oversubtracting" strategy.)

Use place-value discs to show how easy it is to subtract 100 from a number:

- Make the number 340 with discs.
- Take away one 100-disc.
- Point out to students we've taken away 100, which is one more than 99. We've taken away one too many. Ask what we can do to fix this. (Add 1)
- Add one 1-disc.

Give students some three-digit numbers and have them subtract 97, 98, or 99 by subtracting a 100-disc and adding back the correct amount of 1-discs.

20   Teacher's Guide 2B Chapter 8   © 2017 Singapore Math Inc.

Do

1 – 2 These problems encourage students to use a specific strategy.

3 Allow students to use the strategy that works best for them.

Activity

▲ **Subtract 97, 98, or 99**

Materials: Three 10-sided dice or decks of Number Cards (BLM) 0 to 9, die modified with sides labeled: 97, 97, 98, 98, 99, 99 as used in Lesson 4

Each player tries to make the greatest difference on a given round. Players take turn rolling the dice. On each roll, they make a three-digit number from the 10-sided dice and then subtract 97, 98, or 99, depending on what they rolled with the modified die. The player with the greatest difference gets a point. Play 5 rounds.

Then, players play 5 more rounds, trying to make the least difference. The player with the most points after 10 rounds is the winner.

Exercise 8 • page 17

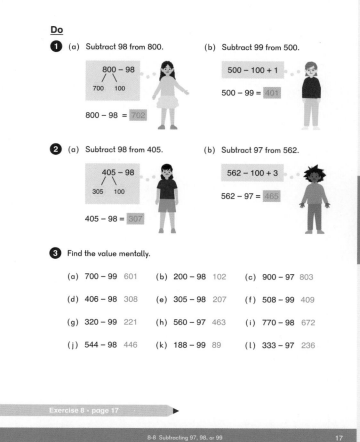

Lesson 9 Practice B

Objective

- Practice subtraction using mental math strategies.

After students complete the **Practice** in the textbook, have them continue to practice mental subtraction with activities from the chapter.

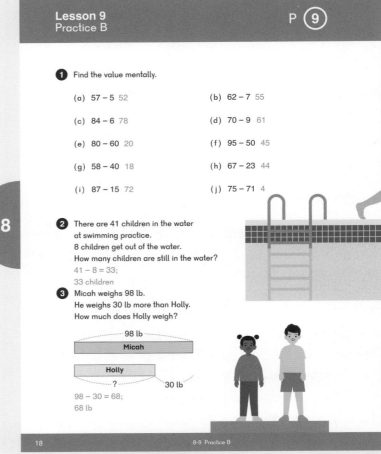

Activity

▲ **Three in a Row – Subtraction**

Materials: Hundred Chart (BLM) in dry erase sleeve, multiple sets of Number Cards (BLM) or playing cards 0 to 9, dry erase markers

Play as described on page 15 of this Teacher's Guide, but subtracting instead.

| 7 | 2 | 1 | 9 |

If a player is dealt the four cards shown above, she can do 27 – 19, which is 8, so she marks 8 on the Hundred Chart (BLM).

Play continues until a player has three marks in a row, column, or diagonal.

If a number is already marked, the digits have to be rearranged to make a different answer.

1	2	3	4	5	6	7	8	9	10
11	12	13	14	15	16	17	18	19	20
21	22	23	24	25	26	27	28	29	30
31	32	33	34	35	36	37	38	39	40
41	42	43	44	45	46	47	48	49	50
51	52	53	54	55	56	57	58	59	60
61	62	63	64	65	66	67	68	69	70
71	72	73	74	75	76	77	78	79	80
81	82	83	84	85	86	87	88	89	90
91	92	93	94	95	96	97	98	99	100

Exercise 9 • page 19

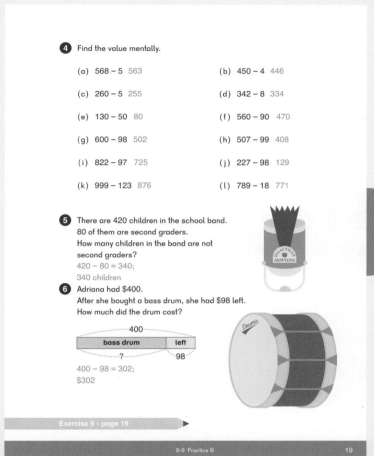

Lesson 10 Practice C

Objective
- Practice addition and subtraction using mental math.

After students complete the **Practice** in the textbook, have them continue to practice mental addition and subtraction with activities from the chapter.

Brain Works

★ Number Puzzles

Materials: Number Cards (BLM) 1 to 9

Puzzle 1: Arrange the digits 1 to 9 into 3 three-digit whole numbers.

Using each digit only once, make the sum as close to 1,000 as possible.

Puzzle 2: Fill in the boxes with digits to make the answer in the problem below closer to 300 than 400.

5 ▢ ▢ − 2 ▢ ▢ =

- What is one solution?
- How many other solutions can you find?

Example answers:
534 − 198, 534 − 197, 534 − 196, 536 − 198...

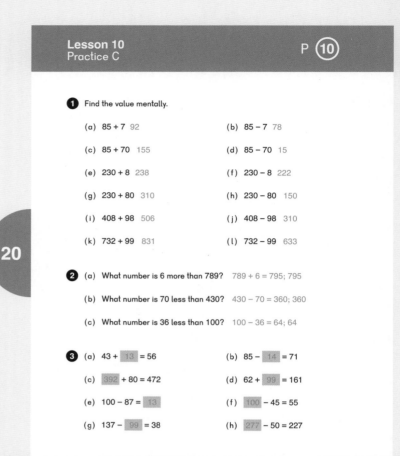

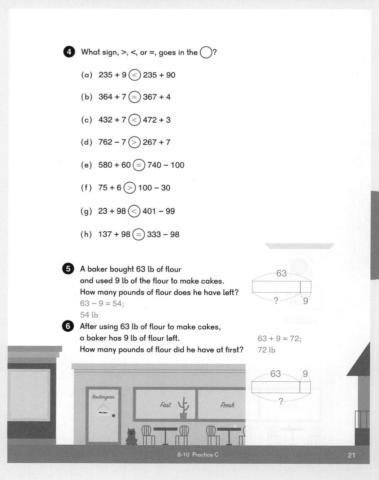

Exercise 10 • page 23

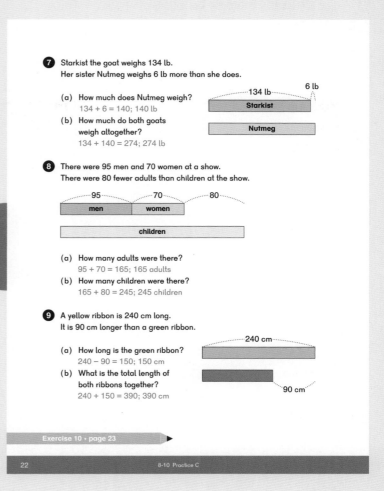

Exercise 1 • pages 1–2

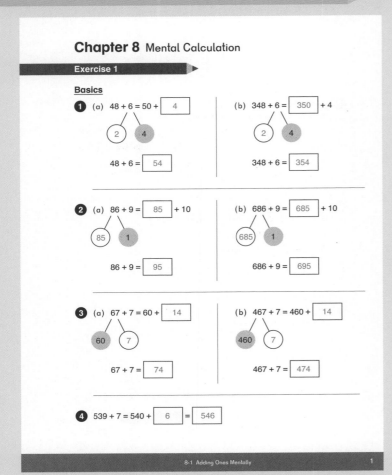

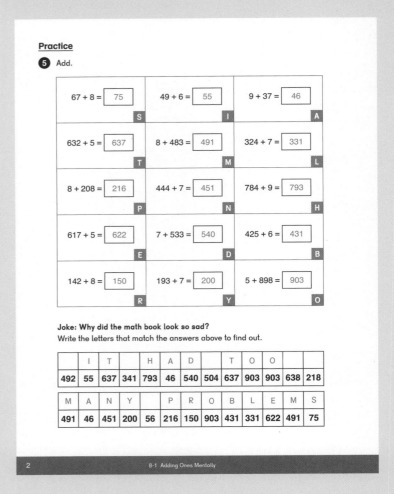

Exercise 2 • pages 3–4

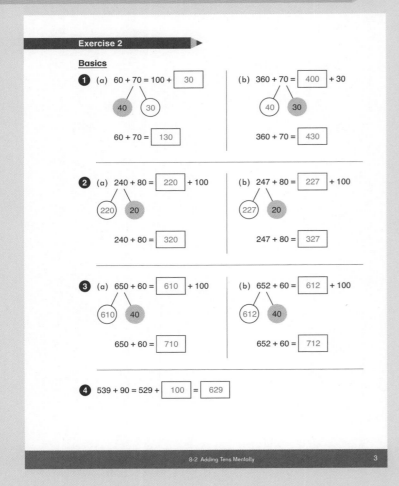

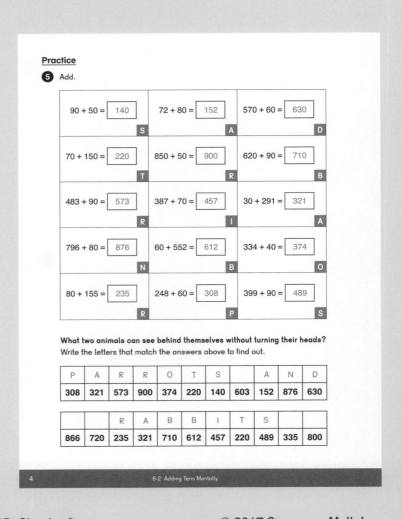

Exercise 3 • pages 5–6

Exercise 3

Basics

1 (a) 9 tens [10] ones = 100

(b) [4] tens 5 ones + 5 tens [5] ones = 9 tens 10 ones

(c) 45 + [55] = 100

2 (a) 60 + [40] = 100 | 100 − 60 = [40]

(b) 64 + [36] = 100 | 100 − 64 = [36]

(c) 64 + [236] = 300 | 300 − 64 = [236]

3 Match pairs of numbers that make 100.

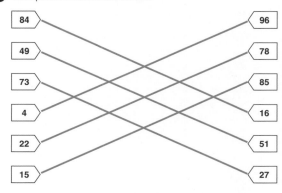

Practice

4 Write the missing numbers.

(a) 74 + [26] = 100 (b) 15 + [85] = 100

(c) 55 + [45] = 100 (d) 81 + [19] = 100

(e) 92 + [8] = 100 (f) [78] + 22 = 100

(g) 100 = [32] + 68 (h) 100 = 3 + [97]

5 Subtract.

(a) 100 − 46 = [54] (b) 100 − 62 = [38]

(c) 100 − 33 = [67] (d) 100 − 57 = [43]

(e) 100 − 75 = [25] (f) [94] = 100 − 6

(g) 100 − [2] = 98 (h) 64 = 100 − [36]

Challenge

6 The numbers on each side of each triangle should add up to 100. Write the missing numbers.

(a) 62 at top; 23, **6** on sides; 15, 53, 32 on bottom

(b) 21 at top; 76, 16 on sides; 3, **34**, 63 on bottom

Exercise 4 • pages 7–8

Exercise 4

Basics

1 (a) 99 + 37 = [100] + 36 (b) 99 + 537 = 100 + [536]
    (1, 36)   (1, 536)
    99 + 37 = [136]   99 + 537 = [636]

2 (a) 85 + 98 = [83] + 100 (b) 485 + 98 = [483] + 100
    (83, 2)   (483, 2)
    85 + 98 = [183]   485 + 98 = [583]

3 73 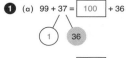 [173] 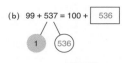 [172]

73  [172]

73 + 99 = [172]

4 (a) 97 = 100 − [3]

(b) 97 + 400 = 500 − [3] = [497]

(c) 97 + 486 = 586 − [3] = [583]

Practice

5 Add.

54 + 99 = 153	38 + 97 = 135	73 + 98 = 171
98 + 97 = 195	99 + 88 = 187	142 + 98 = 240
652 + 99 = 751	325 + 98 = 423	97 + 606 = 703
777 + 98 = 875	98 + 709 = 807	447 + 97 = 544
97 + 119 = 216	324 + 98 = 422	99 + 599 = 698

Color the spaces that contain the answers to help the spider find its home.

204	671	195	153	807	124	806
117	175	135	421	751	432	423
874	875	187	545	544	689	171
216	422	702	973	240	703	698
578	571	214	230	402	161	398

Exercise 5 • pages 9–12

Exercise 5

Check

1 Complete the number patterns.

(a) **138** | 145 | **152** | **159** | 166 | 173 | **180**

(b) **230** | **310** | **390** | 470 | 550 | 630 | 710

(c) **333** | **432** | 531 | 630 | 729 | 828 | **927**

(d) **147** | **244** | 341 | 438 | 535 | 632 | **729**

2 Find the pattern.
Write the missing numbers.

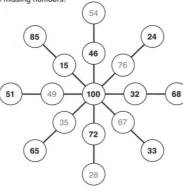

3 Complete the cross-number puzzle using the clues.

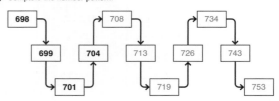

	A		B		C	
	9	4	5		2	
		5		6		7
D 4	2	0		E 6	F 5	9
7					2	
G 4	H 2	3		I 7	3	J 4
		5				7
K 7	0	0			L 6	6	0

Across		Down	
945	A 936 + 9	A 870 + 80	950
420	D 360 + 60	B 468 + 98	566
659	E 619 + 40	C 90 + 189	279
423	G 373 + 50	D 99 + 375	474
734	I 6 + 728	F 453 + 70	523
700	K 660 + 40	H 245 + 5	250
660	L 580 + 80	J 379 + 97	476

4 June is collecting minerals to polish in a rock tumbler.
She has collected 90 rose quartz.
She has 40 more smoky quartz than rose quartz.

(a) How many smoky quartz does she have?

90 + 40 = 130

She has __130__ smoky quartz.

(b) How many of both kinds of quartz does she have in all?

90 + 130 = 220

She has __220__ quartz in all.

5 June had 98 moonstones, and then found another 38 moonstones and 80 sunstones.

(a) How many moonstones does she have in all?

98 + 38 = 136

She has __136__ moonstones in all.

(b) How many moonstones and sunstones does she have in all?

136 + 80 = 216

She has __216__ moonstones and sunstones in all.

Challenge

6 Complete the number pattern.

698 → 699 → **701** → **704** → 708 → 713 → **719** → 726 → **734** → **743** → **753**

7 Add using mental math.

(a) 63 + 499 = **562** (b) 24 + 598 = **622**

(c) 99 + 799 = **898** (d) 398 + 265 = **663**

8 Mei added 85 and 67 like this:

85 + 67 = 85 + 70 − 3 = 152

Use this strategy to mentally add the following.

(a) 437 + 8 = **445** (b) 95 + 78 = **173**

(c) 59 + 35 = **94** (d) 388 + 68 = **456**

Exercise 6 • pages 13–14

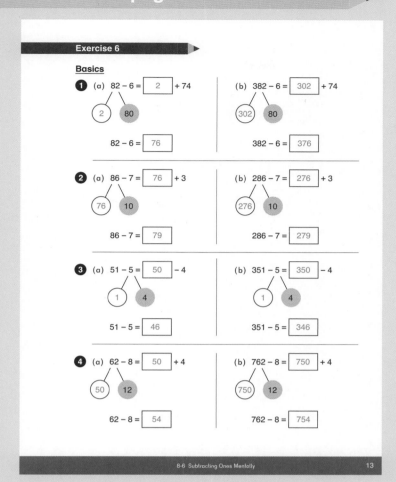

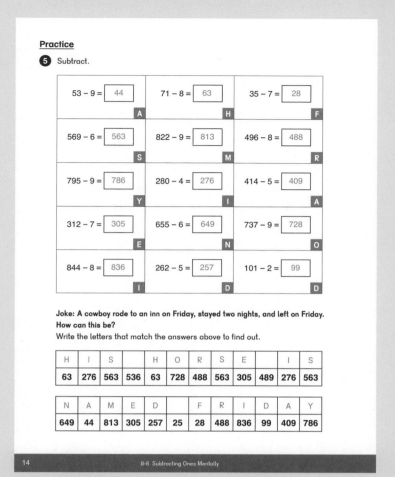

Exercise 7 • pages 15–16

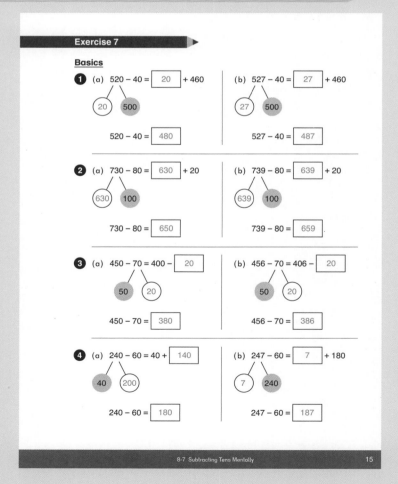

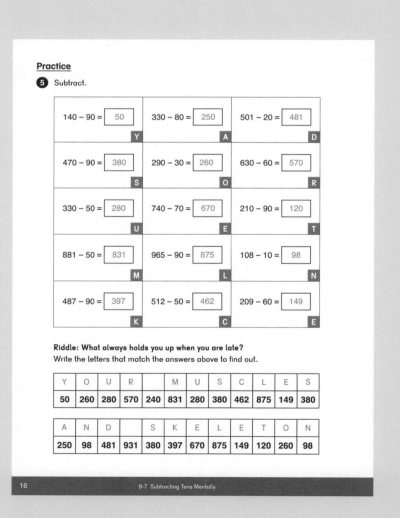

Exercise 8 • pages 17–18

Exercise 8

Basics

1. (a) 300 − 99 = 200 + [1]
   200 100
   300 − 99 = [201]

   (b) 305 − 99 = [205] + 1
   205 100
   305 − 99 = [206]

2. (a) 500 − 98 = 400 + [2]
   400 100
   500 − 98 = [402]

   (b) 585 − 98 = [485] + 2
   485 100
   585 − 98 = [487]

3. 173 —−100→ [73] —+1→ [74]
   173 —−99→ [74]
   173 − 99 = [74]

4. (a) 100 = 97 + [3]
   (b) 400 − 97 = 300 + [3] = [303]
   (c) 486 − 97 = 386 + [3] = [389]

Practice

5. Subtract.

154 − 99 = [55]	600 − 97 = [503]	173 − 98 = [75]
796 − 97 = [699]	301 − 99 = [202]	142 − 98 = [44]
652 − 99 = [553]	325 − 98 = [227]	891 − 97 = [794]
777 − 98 = [679]	313 − 97 = [216]	443 − 98 = [345]
734 − 97 = [637]	324 − 98 = [226]	998 − 99 = [899]

Riddle: What is at the end of a rainbow?
Color the spaces that contain the answers to find out.

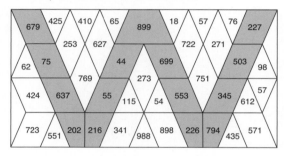

Exercise 9 • pages 19–22

Exercise 9

Check

1 Subtract.

410 − 70 = **340** U	500 − 98 = **402** A	100 − 62 = **38** Y
800 − 30 = **770** T	120 − 50 = **70** M	632 − 7 = **625** E
546 − 50 = **496** H	740 − 97 = **643** C	210 − 90 = **120** O
333 − 40 = **293** E	525 − 9 = **516** N	632 − 98 = **534** R

Joke: What did one wall say to the other?
Write the letters that match the answers above to find out.

M	E	E	T		Y	O	U		A	T	
70	625	293	770	596	38	120	340	210	402	770	562

T	H	E		C	O	R	N	E	R
48	770	496	625	637	643	120	534	516	293	534	435

2 Complete the cross-number puzzle using the clues.

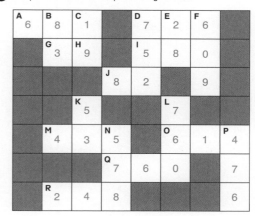

Across
- 681 **A** 741 − 60
- 726 **D** 733 − 7
- 39   **G** 137 − 98
- 580 **I** 620 − 40
- 82   **J** 91 − 9
- 435 **M** 495 − 60
- 614 **O** 712 − 98
- 760 **Q** 830 − 70
- 248 **R** 268 − 20

Down
- **B** 100 − 17   83
- **C** 116 − 97   19
- **D** 760 − 8   752
- **E** 100 − 72   28
- **F** 614 − 5   609
- **K** 151 − 98   53
- **L** 840 − 80   760
- **N** 582 − 4   578
- **P** 506 − 30   476

3 June made 132 earrings and 80 necklaces last month using polished rocks.

(a) How many more earrings than necklaces did she make?

132 − 80 = 52

She made ___52___ more earrings than necklaces.

(b) June sold 98 of the earrings.
How many earrings did she have left?

132 − 98 = 34

She has ___34___ earrings left.

4 June made $350 from selling jewelry.
She wants to make $410.

(a) How much more money does she need to make?

410 − 350 = 60

She needs to make $___60___ more.

(b) She spent $90 of the $350 she made on more craft items to make more jewelry.
How much money does she have left of that $350?

350 − 90 = 260

She has $___260___ left.

Challenge

5 Complete the number pattern.

626 → **616** → **596** → **526** → **566** → **476** → **416** → **266** → **346** → **176** → **76**

6 Subtract using mental math.

(a) 245 − 199 = **46**   (b) 963 − 499 = **464**

(c) 391 − 297 = **94**   (d) 924 − 598 = **326**

7 Dion subtracted 68 from 453 like this:

453 − 68 = 453 − 70 + 2 = 385

Use the strategy to mentally subtract the following.

(a) 437 − 8 = **429**   (b) 46 − 29 = **17**

(c) 59 − 35 = **24**   (d) 324 − 48 = **276**

© 2017 Singapore Math Inc.   Teacher's Guide 2B Chapter 8

Exercise 10 • pages 23–26

Exercise 10

Check

1 Find the values.

(a) 237 + 98 = 335 (b) 237 − 98 = 139
(c) 64 + 99 = 163   (d) 132 − 98 = 34
(e) 373 − 97 = 276 (f) 425 + 98 = 523
(g) 278 − 99 = 179 (h) 631 + 97 = 728
(i) 843 + 99 = 942 (j) 555 − 98 = 457

2 Write >, <, or = in each ◯.

(a) 100 − 65 < 100 − 56
(b) 42 + 612 = 642 + 12
(c) 82 + 60 > 142 − 50
(d) 147 − 99 < 47 + 99
(e) 375 + 8 < 317 + 80
(f) 630 − 80 = 480 + 70
(g) 132 + 7 + 40 > 124 + 7 + 30
(h) 800 + 40 + 50 > 900 − 90

3 Follow the arrows and fill in the missing numbers.

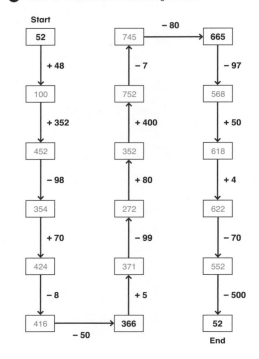

4 There are 206 bones in an adult human skeleton.
30 of them are in the spine.
How many bones are not in the spine? **176**

206 − 30 = 176

5 Dexter found a snake skeleton and counted 310 vertebrae.
Each vertebra had 2 ribs attached.

(a) How many ribs did the snake skeleton have? **620**

310 + 310 = 620

(b) How many bones did it have for both vertebrae and ribs? **930**

620 + 310 = 930

(c) The head of the snake had 10 bones.
How many bones did the snake have in all? **940**

930 + 10 = 940

(d) How many more bones did this snake have than an adult human skeleton? **734**

940 − 206 = 734

Challenge

6 Find the number that each shape stands for.

◆ + ⬡ = 100   ◆ = 63   Hint: Find ◆ first.
◆ − ⬡ = 26   ⬡ = 37
26 + ◆ = 89

7 The numbers on each circle should add up to the number in the square between them.
Write the missing numbers.

(a) 80; 150, 130; 70, **120**, 50
(b) 10; 130, 190; 120, **300**, 180

8 There are some bees and flowers.
If each bee lands on a different flower, one bee does not get a flower.
If two bees share each flower, there is one flower left out.
How many flowers and bees are there?
4 bees 3 flowers
Students can draw a picture, starting with two bees and two flowers, and then experiment with what happens if a bee or a flower is added.

Teacher's Guide 2B Chapter 8

Chapter 9 Multiplication and Division of 3 and 4 — Overview

Suggested number of class periods: 9–10

	Lesson	Page	Resources	Objectives
	Chapter Opener	p. 37	TB: p. 23	Investigate multiplication and division of 3 and 4.
1	The Multiplication Table of 3	p. 38	TB: p. 24 WB: p. 27	Build and understand the structure of the multiplication table of 3. Look for patterns in the multiplication table of 3.
2	Multiplication Facts of 3	p. 41	TB: p. 27 WB: p. 29	Understand the commutative property in the multiplication table of 3. Learn the multiplication facts of 3.
3	Dividing by 3	p. 44	TB: p. 30 WB: p. 31	Use a related multiplication sentence to solve division problems without a remainder where the divisor is 3.
4	Practice A	p. 47	TB: p. 33 WB: p. 35	Practice the multiplication and division facts of 3. Solve problems involving multiplication and division by 3.
5	The Multiplication Table of 4	p. 48	TB: p. 34 WB: p. 39	Build and understand the structure of the multiplication table of 4. Look for patterns in the multiplication table of 4.
6	Multiplication Facts of 4	p. 51	TB: p. 37 WB: p. 41	Understand the commutative property in the multiplication table of 4. Learn the multiplication facts of 4.
7	Dividing by 4	p. 53	TB: p. 40 WB: p. 43	Use a related multiplication sentence to solve division problems without a remainder where the divisor is 4.
8	Practice B	p. 55	TB: p. 42 WB: p. 47	Practice the multiplication and division facts of 4. Solve problems involving multiplication and division by 4.
9	Practice C	p. 56	TB: p. 44 WB: p. 51	Practice multiplying and dividing by 2, 3, 4, 5, and 10. Solve two-step word problems involving multiplying and dividing by 2, 3, 4, 5, and 10.
	Workbook Solutions	p. 60		

Chapter 9 Multiplication and Division of 3 and 4

In this chapter, students formalize their knowledge of multiplication and division for 3 and 4.

Lesson 1 introduces building multiplication tables for 3 with activities similar to those in **2A** Chapter 7: Multiplication Tables of 2, 5, and 10. Lesson 2 will continue to develop and practice the facts, and will reiterate that the answer is the same regardless of which number comes first, the number of groups or the number in each group.

In Lesson 3, students learn that they can find their division facts by thinking of a related multiplication fact. The lesson also reiterates the two ideas of division, sharing and grouping, and that it does not matter if the divisor represents the number of groups or the number in each group, the quotient (which will be the number in each group or the number of groups respectively) is the same.

Lessons 5, 6, and 7 will follow the same structure to explore multiplication by 4. Most games and activities from this chapter can be adapted for different numbers in the multiplication and division tables. They should be used throughout the year.

Students should continue to practice these facts until they know them from memory.

Students will learn the remaining 16 facts for multiplication and division by 6, 7, 8, and 9 in **Dimensions Math® 3A**.

Bar Models

Teachers interested in using multiplication and division bar models should reference **Dimensions Math® 2A** Chapter 7. At this level, however, it is more beneficial for students to act out the word problems with objects or make drawings than it is to use bar models.

x	1	2	3	4	5	6	7	8	9	10
1	1	2	3	4	5	6	7	8	9	10
2	2	4	6	8	10	12	14	16	18	20
3	3	6	8	12	15	18	21	24	27	30
4	4	8	12	16	20	24	28	32	36	40
5	5	10	15	20	25	30	35	40	45	50
6	6	12	18	24	30					60
7	7	14	21	28	35					70
8	8	16	24	32	40					80
9	9	18	27	36	45					90
10	10	20	30	40	50	60	70	80	90	100

Chapter 9 Multiplication and Division of 3 and 4 — Materials

Materials

- Construction paper
- Counters
- Craft punches/hole punches
- Dot stickers
- Dry erase markers
- Dry erase sleeves
- Fly swatters or bean bags
- Glue
- Index cards
- Markers or crayons
- Modified die with sides labeled: 2, 3, 4, 4, 5, 10
- Paper
- Paper plates
- Pebble
- Recording sheet
- Sidewalk chalk or painter's tape
- Sticky notes
- Whiteboards

Blackline Masters

- Array Dot Cards — 3
- Array Dot Cards — 4
- Individual Number Path
- Kaboom Cards
- Multiplication and Division Fact Cards for 2
- Multiplication and Division Fact Cards for 3
- Multiplication and Division Fact Cards for 4
- Multiplication and Division Fact Cards for 5
- Multiplication and Division Fact Cards for 10
- Multiplication Chart 9-1
- Multiplication Chart 9-5
- Secret Math Message Decoder

Storybooks

- *Math Attack!* by Joan Horton
- *2 × 2 = Boo!* by Loreen Leedy
- *Bean Thirteen* by Matthew McElligot
- *Divide and Ride* by Stuart J. Murphy
- *Spunky Monkeys on Parade* by Stuart J. Murphy
- *Too Many Kangaroo Things to Do!* by Stuart J. Murphy

Notes

Chapter Opener

Objective

- Investigate multiplication and division of 3 and 4.

Lesson Materials

- Counters

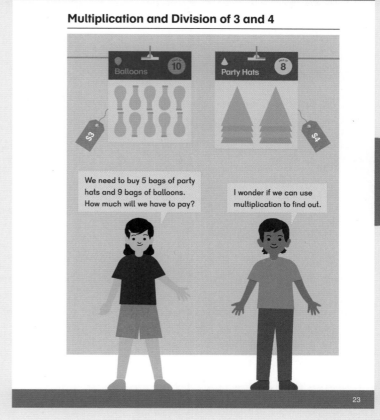

Review facts of 2, 5, and 10 as a warm-up and play games from Chapter 7: Multiplication and Division of 2, 5, and 10.

Provide students with counters, if necessary, and have them find the price of the hats and balloons.

Have students share some multiplication or division equations based on what they know about multiplication facts of 2, 5, and 10.

Lesson 1 The Multiplication Table of 3

Objectives

- Build and understand the structure of the multiplication table of 3.
- Look for patterns in the multiplication table of 3.

Lesson Materials

- Counters, 30 per student
- Multiplication Chart 9—1 (BLM)
- Array Dot Cards — 3 (BLM)

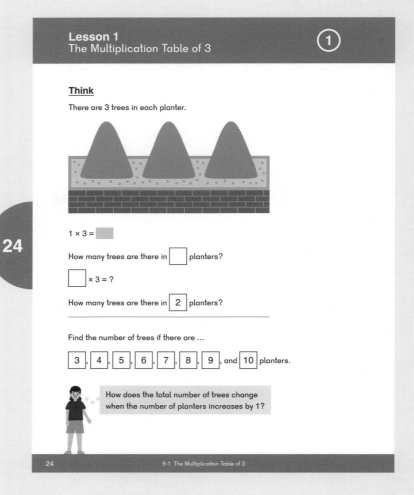

Think

Pose the **Think** problem. Ask students what they have learned this year that might help them with this problem.

Have students use counters and share how they solved the problem.

As in the **Dimensions Math® 2A** lessons for multiplication by 2, 5, and 10, students will build and record the multiplication table for 3.

Have students start by making one row of counters and filling in Multiplication Chart 9—1 (BLM):

- 1 package, 3 glue sticks in each package, total number of glue sticks. The product column should be left blank.

Have students add a second row of 3 counters to see how many glue sticks there are in 2 packages.

Continue to add rows and complete the chart. Reinforce the term **array** when laying out the rows.

Discuss Mei's thought on the increasing total.

Ask students:

- How is this problem different from the problems in **Dimensions Math® 2A** when we counted mangoes by fives, watermelons by twos, and walnuts by tens? (We are multiplying 3, or by 3.)
- How is it the same?
  (We are still making equal groups.)

Learn

Have students add the equations to their Multiplication Chart 9–1 (BLM).

Students should notice that as the number of groups increases by 1, the product increases by 3. If students know 3 × 3, they can find 4 × 3 by simply adding 3, rather than counting by threes from the start.

Students should also notice that 9 × 3 is the same as 10 × 3 − 3, or 30 − 3.

Discuss Alex's comments. Students will be introduced to factors, or multiples, in **Dimensions Math® 4A**. At this point, accept any reasonable answer.

When you add the digits of the sum of the products of 3, the sums:

- Create a pattern: 3, 6, 9, 3, 6, 9…
- Are also multiples of 3.

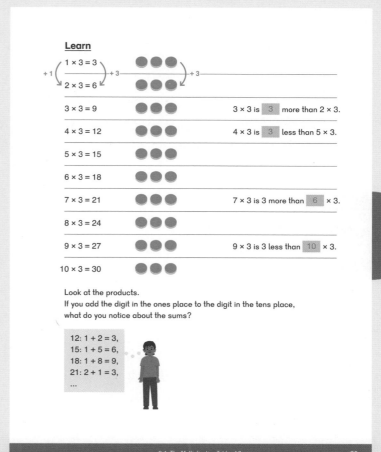

Do

1 Students may notice the pattern that 4 is double 2, so 4 × 3 is double 2 × 3.

3 Do the activity below and have students create their own dot cards. Alternatively, students can use Array Dot Cards — 3 (BLM) in dry erase sleeves for the problems.

Activity

▲ Array Dot Cards — 3

Materials: Index cards, craft punches/hole punches or dot stickers, glue

Have each student fold a piece of paper into 8 parts then cut along the lines to create 8 cards.

Have each student use either dot stickers or punches to make a 3 × 10 array on index cards for future reference. Students will make similar cards for × 4.

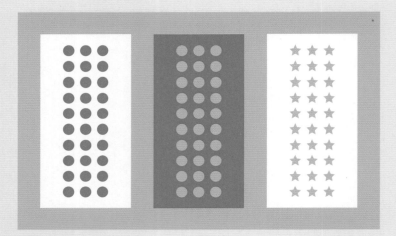

Exercise 1 • page 27

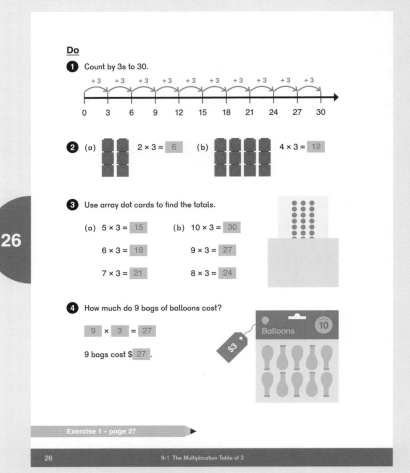

Lesson 2 Multiplication Facts of 3

Objectives

- Understand the commutative property in the multiplication table of 3.
- Learn the multiplication facts of 3.

Lesson Materials

- Counters
- Construction paper or index cards, 20 per student

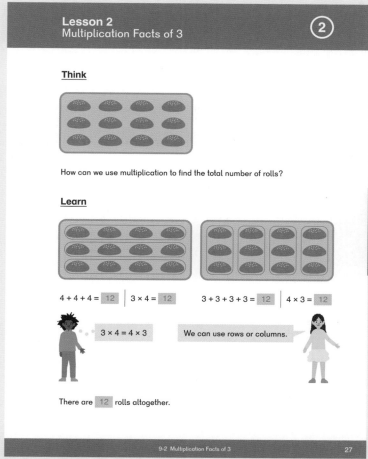

Think

Pose the **Think** problem. At this point, most students will know that there are 12 rolls. Students may mention that the rolls are arranged in an array.

Have students model the problem with counters and share how they solved the problem.

Learn

Dion reminds us that the order in which we multiply will not change the total. Whether it is 3 groups of 5 or 5 groups of 3, the total is the same.

While learning the facts, the commutative property is very valuable. When students learn 3 × 6, they know the answer will be the same as for 6 × 3, which they have already learned. Thus, they know some facts for groups of 6 before they count by sixes.

This means there are fewer and fewer facts that need to be learned in each successive table.

This part of the lesson may progress quickly. Students will need time in the **Do** part of the lesson to create flash cards.

© 2017 Singapore Math Inc.   Teacher's Guide 2B Chapter 9   41

Do

1 – 2 Emphasize that it does not matter which number comes first, the number of groups or the number in each group, the answer is the same. The numbers can be multiplied in any order to get an answer.

3 Students should see that 3 grams × 6 is equal to 6 grams × 3.

5 Provide students with index cards and have them create their own flash cards for future practice and games.

Students can also fold construction paper into 8 equal parts and cut out their own flash cards.

Activities

▲ **Multiplication Wheels**

Materials: Paper plates with the center cut out

Create several multiplication wheels with the numbers 1 to 10 in random order as "spokes" along the edge of the paper plates.

Students lay the wheel on a whiteboard and write the number they are multiplying by in the center of the wheel. (In this lesson it is 3.)

Students multiply the number on the spoke and the number in the center, and write the product on the whiteboard, outside of the wheel.

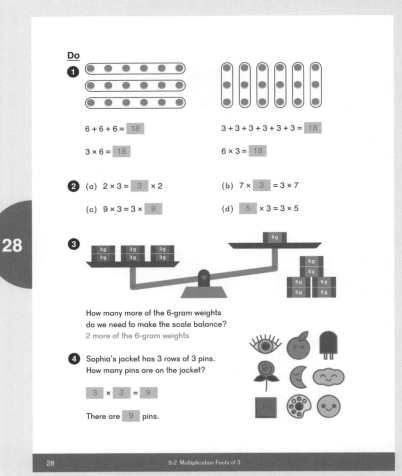

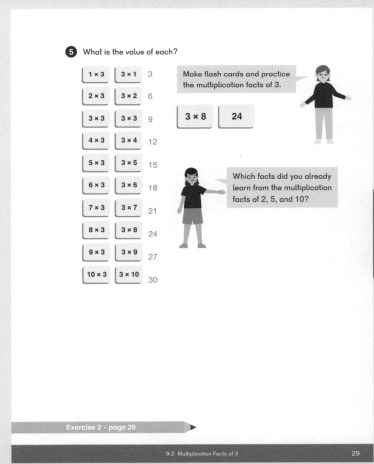

▲ **Three's a Hopping!**

Materials: Sidewalk chalk or painter's tape, multiplication by 3 cards made in this lesson

Create two grids like the one shown below, using either chalk outside or painter's tape inside.

3	12	27
18	6	9
24	15	21
30	17	Home

One student is the Caller. Two students are the Hoppers, and stand on their home squares. The Caller flips over a multiplication by 3 card and calls out the equation.

Hoppers must hop on the answer.

The first Hopper who misses the correct square becomes the next Caller. (Include a non-multiple of 3 in the extra square.)

▲ **Three's a Bopping!**

Materials: Paper, markers or crayons, fly swatters or bean bags, multiplication by 3 cards made in this lesson

Create two grids on paper like the one shown above to play a smaller version of **Three's a Hopping** inside.

In this version, players either smack the numbers with a fly swatter or toss a bean bag onto the answer.

Exercise 2 • page 29

Lesson 3 Dividing by 3

Objective

- Use a related multiplication sentence to solve division problems without a remainder where the divisor is 3.

Lesson Materials

- Counters
- Paper plates

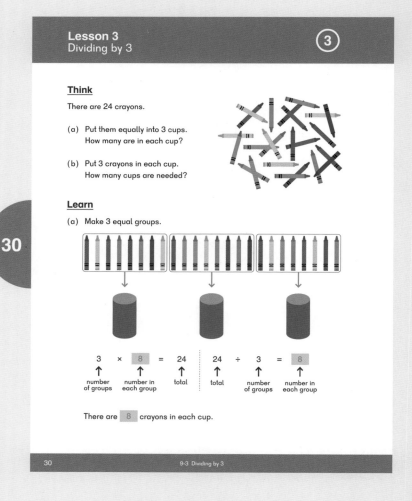

Think

Pose the **Think** problem. Provide students with counters and paper plates to work through the problem using the plates to show the equal groups.

Have students share their strategies.

Learn

Discuss the examples in the textbook. Ask students if each is a sharing or grouping problem. The first example shows equal groups and students will find how many crayons there are in a group.

Problem (b) and example (b) show groups of 3 crayons. Students will find how many groups there are in all.

Have students discuss how the two situations are different. In one we are finding the number in each group (sharing), in the other, we are finding the number of groups (grouping).

Encourage students to use the language of division.

- The crayons are divided into 3 groups.
- The crayons are divided into groups of 3.

Alex points out that if students know their multiplication facts, they can use them to find division facts.

Students can think about division by using multiplication:

$3 \times ? = 24$

$? \times 3 = 24$

Ensure students understand that the answer is the same.

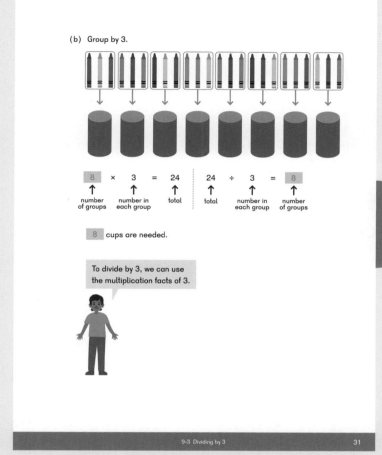

Do

① Have students discuss the different situations, and then show the two situations with a drawing.

In (a) there are 3 groups, however, in (b) there are groups of 3.

③ Have students share a related multiplication equation.

Activities

▲ **Clear the Board Division**

Materials: Division by 3 Fact Cards (BLM), Individual Number Path (BLM), and 7 counters for each player

Give each student an Individual Number Path (BLM), and have them put counters on 7 of the numbers.

Players take turns drawing Division by 3 Fact Cards (BLM). If the answer to the card a player draws is one of the numbers covered on her Individual Number Path (BLM), she removes the counter from that number. The first player to clear all of her counters is the winner.

▲ **Divide by 3 Kaboom**

Materials: Kaboom Cards (BLM), several sets of Division by 3 Fact Cards (BLM)

Shuffle and place the Division by 3 Fact Cards (BLM) facedown in a pile. Players take turns drawing a card and saying the answer to the division fact.

They keep the cards they answer correctly, and return the cards they answer incorrectly back to the pile.

When a student draws a Kaboom Card (BLM), he must return all his collected cards to the pile.

The player with the most cards at the end of the time limit is the winner.

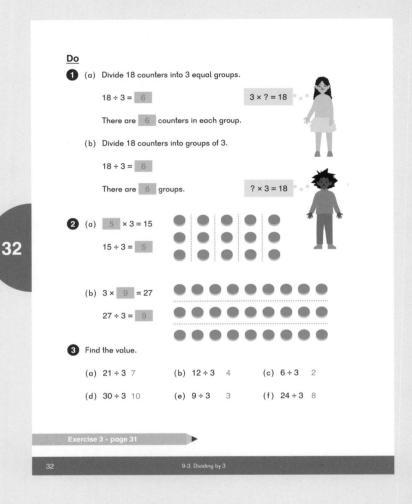

Exercise 3 • page 31

Lesson 4 Practice A

Objectives

- Practice the multiplication and division facts of 3.
- Solve problems involving multiplication and division by 3.

After students complete the **Practice** in the textbook, have them continue to practice multiplication facts for 3 and review facts for 2, 5, and 10 with activities and flash cards from the chapter.

Students should continue to practice these facts until they know them from memory.

Activity

★ Up or Out

Materials: Recording sheet for each player, Multiplication and Division Facts for 2 (BLM), Multiplication and Division Facts for 3 (BLM), Multiplication and Division Facts for 5 (BLM), Multiplication and Division Facts for 10 (BLM)

Shuffle the cards and place them facedown in a pile. Players each begin with 100 points on their recording sheet.

On each turn, a player draws a card and solves the problem. If the card is a multiplication fact, the player adds the product to her total.

If the card is a division fact, the player subtracts the answer from her total. The first player whose total is greater than 300 or less than 10 is the winner.

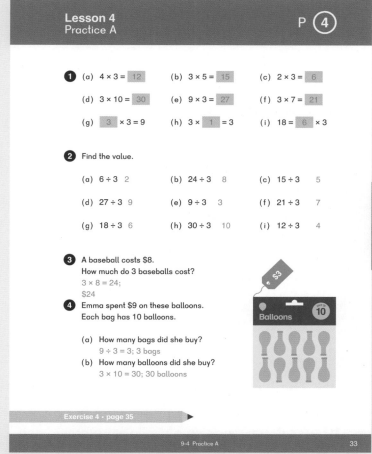

Exercise 4 • page 35

© 2017 Singapore Math Inc.   Teacher's Guide 2B Chapter 9   47

Lesson 5 The Multiplication Table of 4

Objectives

- Build and understand the structure of the multiplication table of 4.
- Look for patterns in the multiplication table of 4.

Lesson Materials

- Counters, 34 per student
- Multiplication Chart 9—5 (BLM)
- Index cards
- Craft punches/hole punches or dot stickers
- Glue
- Array Dot Cards — 4 (BLM)
- Dry erase sleeves

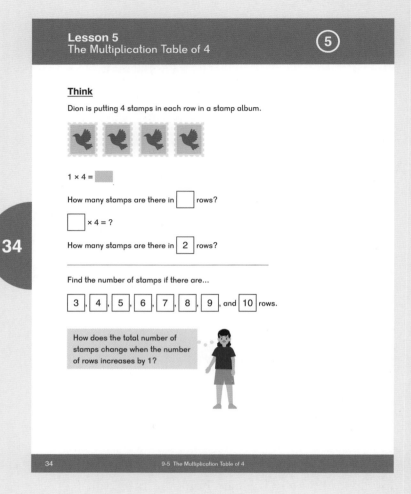

Think

Pose the **Think** problem with the stamps. Ask students if they recall similar problems with 2, 3, 5, and 10. Ask if they can use a similar strategy to solve these problems.

Have students use counters to show how they found the number of stamps.

As in the lessons for multiplication by 2, 5, 10, and 3, students will build and record the multiplication table for 4 using Multiplication Chart 9—5 (BLM).

Learn

Discuss Mei's thought on the increasing total.

Ask students:

- How is this problem different from the problems in **Dimensions Math® 2A** when we counted mangoes by fives, watermelons by twos, and walnuts by tens? (We are multiplying 4, or by 4.)
- How is it the same?
 (We are still making equal groups.)

Have students continue to analyze the charts they created. They can add the equations to their charts.

Students should notice that as the number of groups increases by 1, the product increases by 4.

Discuss Alex's thoughts. Students may notice that the fours facts are double the twos facts.

Students should also notice that 9 × 4 is the same as 10 × 4 − 4, or 40 − 4. Or, as with all multiplication by 4, as the number of groups decreases by 1, the product decreases by 4.

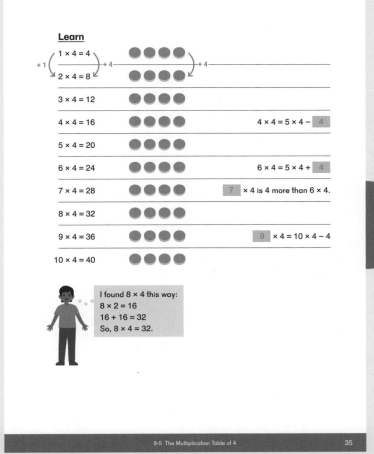

Do

① Students may notice the pattern that 4 is double 2, so 4 × 4 is double 2 × 4.

③ See the Lesson 1 activity **Array Dot Cards** and have students create dot cards for the fours table. Alternatively, students can use Array Dot Cards — 4 (BLM) in dry erase sleeves for the problems.

Alternatively, use dot paper in dry erase sleeves for the problems.

Activities

▲ Array Dot Cards — 4

Materials: Index cards, craft punches/hole punches or dot stickers, glue

Repeat the activity from Lesson 1 on page 40, but this time have students create 4 × 10 arrays for future use.

▲ Hopscotch × 4

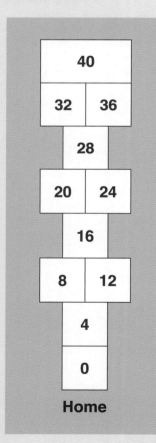

Materials: Pebble or other marker for each player, sidewalk chalk or paper plates and painter's tape

Play outside or in a gym. Draw a hopscotch board using chalk, or tape down paper plates to create a hopscotch board with 0 as the starting spot and the numbers in the squares increasing by 4.

Players take turns standing in the 0 square (Home) and tossing the marker. On their first turn, players aim for the 4 square. On each turn, players hop over the square with the marker and

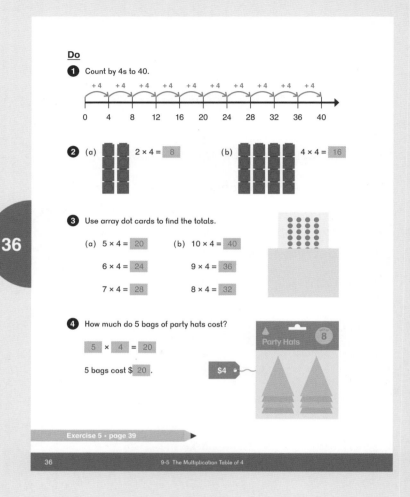

continue hopping in order, saying the numbers in each square aloud.

Square 40 is a rest stop. Players can put both feet down before turning around and hopping back to 0. Players pause in square 8 to pick up the marker, hop in square 4, and out again. On his next turn, the player aims his marker for square 8, etc.

A player's turn is over if:

- His marker does not land in the correct square.
- He loses his balance and puts a second foot or a hand down.
- He lands in a square where the marker is.

The winner is the first player to get through all 10 turns.

◀ **Exercise 5 • page 39**

Lesson 6 Multiplication Facts of 4

Objectives

- Understand the commutative property in the multiplication table of 4.
- Learn the multiplication facts of 4.

Lesson Materials

- Counters
- Construction paper or index cards, 20 per student

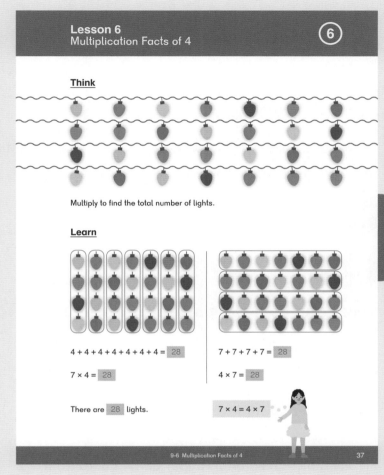

Think

Pose the **Think** problem. At this point, most students will know that there are 28 lights. Students may mention that the lights are arranged in an array.

Have students model the problem with counters and share their solutions.

Learn

Sofia reminds us that the order in which students multiply will not change the total. Whether it is 4 groups of 6 or 6 groups of 4, the total is the same.

While learning the facts, the commutative property will become very valuable. When students learn 4 × 8, for example, they know the answer will be the same as for 8 × 4, which they have already learned. Thus, they know some facts for groups of 8 before they count by eights.

This means there are fewer and fewer facts that need to be learned in each successive table.

As with threes, this part of the lesson may progress quickly. Students will need time in the **Do** part of the lesson to create flash cards.

Do

① – ② Emphasize that it does not matter which number represents the number of groups and which the number in each group, the answer is the same. The numbers can be multiplied in any order to get an answer.

⑤ Provide students with index cards and have them create their own flash cards for future practice and games.

Students can also fold construction paper into 8 equal parts and cut out their own flash cards.

Activity

▲ **Three and Four Hopping and Bopping!**

Materials: Chalk or painter's tape and paper plates, paper and markers or crayons, Multiplication Fact Cards for 3 (BLM) (except 1 × 3), Multiplication Fact Cards for 4 (BLM) (except 1 × 4), fly swatters or bean bags

Modify the chalked or on-paper game board from Lesson 2 page 43 of this Teacher's Guide using fact cards for threes and fours.

Note that the 24 square will work for 3 × 8 and 4 × 6.

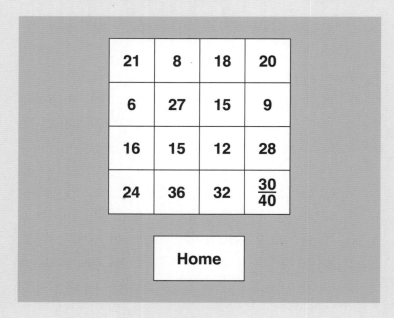

Exercise 6 • page 41

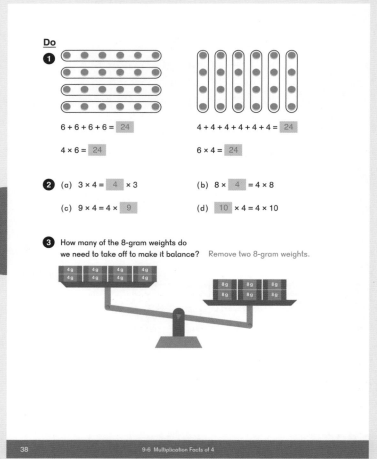

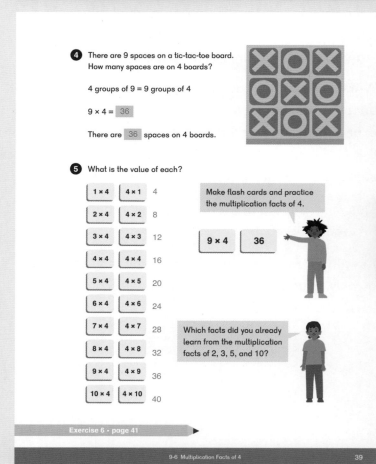

Lesson 7 Dividing by 4

Objective
- Use a related multiplication sentence to solve division problems without a remainder where the divisor is 4.

Lesson Materials
- Counters
- Paper plates

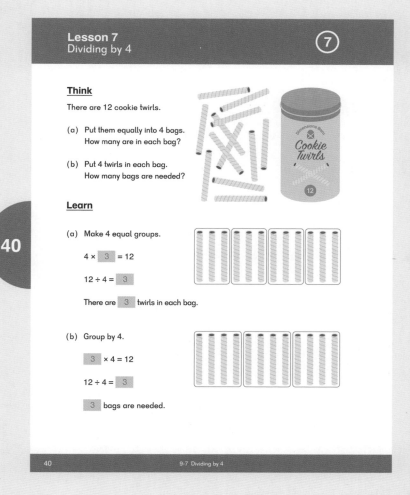

Think

Pose the **Think** problem. Provide students with counters and paper plates to work through the problem.

Discuss student strategies for solving the problems.

Learn

Ask students if these are grouping or sharing situations. The first example shows equal groups and students will find how many cookie twirls there are in a bag.

Problem (b) and example (b) show groups of 4 cookie twirls and students will find how many bags are needed to hold all the cookie twirls.

Have students discuss the two situations. In one, we are finding the number in each group (sharing), in the other, the number of groups (grouping).

Encourage students to use the language of division.

- The cookie twirls are divided into 4 groups.
- The cookie twirls are divided into groups of 4.

© 2017 Singapore Math Inc.   Teacher's Guide 2B Chapter 9   53

Do

1. Have students discuss the different situations. Have them show the two situations with a drawing.

   In (a) there are 5 groups of 4 counters each, in (b) there are 5 counters in each of 4 groups.

3. Have students share a related multiplication equation.

Activities

▲ **Clear the Board Division**

Materials: Division by 4 Fact Cards (BLM), Individual Number Path (BLM), and 7 counters for each player

Play as directed in Lesson 3 on page 46 of this Teacher's Guide.

▲ **Divide by 4 Kaboom**

Materials: Kaboom Cards (BLM), several sets of Division by 4 Fact Cards (BLM)

Play as directed in Lesson 3 on page 46 of this Teacher's Guide.

Exercise 7 • page 43

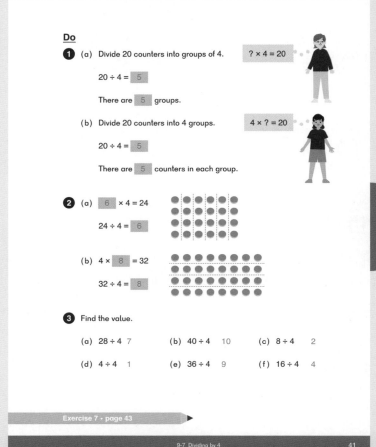

Lesson 8 Practice B

Objectives

- Practice the multiplication and division facts of 4.
- Solve problems involving multiplication and division by 4.

Activity

★ Multiply Team Race

Students seeking a challenge will enjoy the race element of this game.

Have 2–4 students go to the board and make a large "T" shape. Randomly call out 5 numbers from 1 to 10 and have students write them on the left side of the T.

Then give them a number to multiply by, for example, "times 4" in this scenario.

Students write "× 4" on the top of the T, then proceed to solve the problems on the right side as fast as they can.

The first person to complete their problems gets a point for their team.

Exercise 8 • page 47

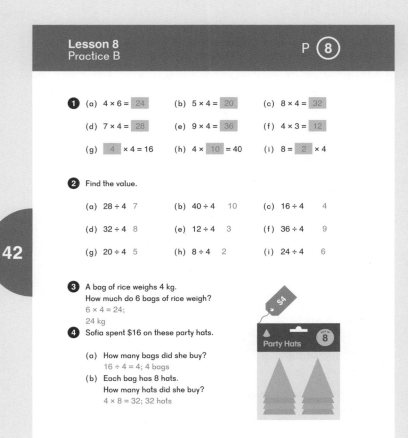

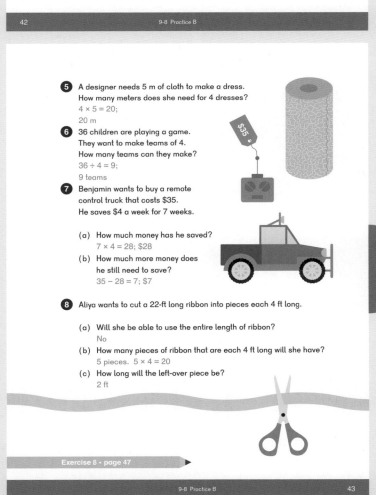

Lesson 9 Practice C

Objectives

- Practice multiplying and dividing by 2, 3, 4, 5, and 10.
- Solve two-step word problems involving multiplying and dividing by 2, 3, 4, 5, and 10.

Materials

- Sticky notes

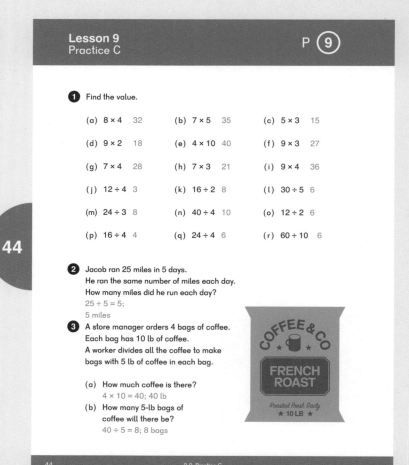

After students complete the **Practice** in the textbook, have them continue to practice multiplication facts for 2, 3, 4, 5, and 10.

Students should continue to practice these facts until they know them from memory. Fluency with these facts will be needed to develop understanding of multi-digit algorithms in **Dimensions Math® 3A**.

5. Remind students that we need to find the missing number that will make these equations true. For example, if 5 × 4 = 20, what number × 2 also equals 20?

6 – 13. The strategy introduced in **Dimensions Math® Teacher's Guide 2A** using sticky notes and whiteboards can be used for these problems.

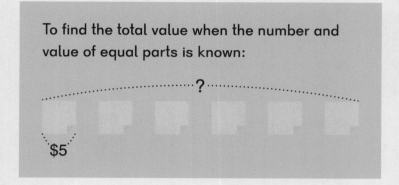

To find the total value when the number and value of equal parts is known:

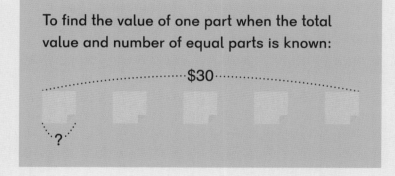

To find the value of one part when the total value and number of equal parts is known:

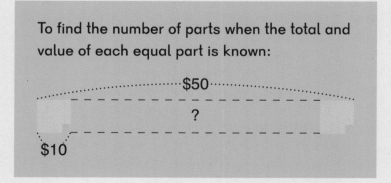

To find the number of parts when the total and value of each equal part is known:

Sticky notes help demonstrate that each bar is an equal amount (or unit) and represents the same amount in a problem. The whiteboard allows multiple problems to be solved by rearranging the bars (sticky notes), numbers, and question mark.

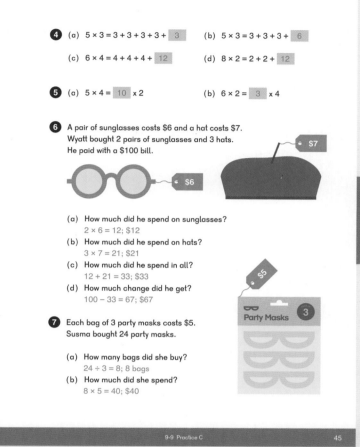

Do

9 – 13 These are two-step problems where the first step is not given. Questions to ask about the problems:

- Is a total given?
- Do we know the parts?
- Do we know a part and a total?
- What information do we know?

10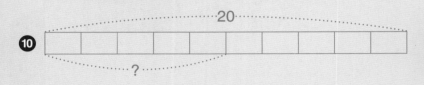

11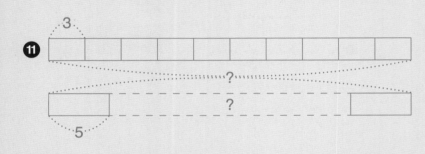

13

8 Maurice spent $15 on balloons.
There are 10 balloons in each bag.
He used 35 balloons at a party.
   $15 \div 3 = 5$; 5 bags; $5 \times 10 = 50$; 50 balloons
   (a) How many balloons did he buy?
   (b) How many balloons did he have left?
      $50 - 35 = 15$; 15 balloons

9 How much do 5 bags of party hats and 9 bags of balloons cost?
   $5 \times 4 = 20$; $9 \times 3 = 27$;
   $20 + 27 = 47$; $47

10 10 lb of cherries cost $20.
Debra bought 5 lb of cherries.
How much did she spend?
   $20 \div 10 = 2$; $2 \times 5 = 10$; $10
   OR: $20 \div 2 = 10$; $10

11 Ella saved $3 a week for 10 weeks.
She wants to buy gifts that cost $5 each for her friends.
How many gifts can she buy?
   $10 \times 3 = 30$; $30 \div 5 = 6$; 6 gifts

12 1 table can seat 4 people.
What is the fewest number of tables needed to seat 34 people?
   9 tables;
   $8 \times 4 = 32$, $9 \times 4 = 36$. 8 is too few, 9 is enough.

13 Carter made a paper chain for the party.
He put 4 yellow strips between every red strip.
There are 8 red strips in the chain.
How many yellow strips are there?
   $7 \times 4 = 28$; 28 yellow strips
   (7 groups of 4 between 8 red strips)

Exercise 9 • page 51

46   9-9 Practice C

46

Activity

▲ Leftovers

Materials: 45 counters, modified die with sides labeled: 2, 3, 4, 4, 5, 10

Player One rolls the die and divides the counters by the number on the die.

For example, Player One rolls a 2. She divides the counters into 2 equal groups with 1 left over. She keeps the leftover counter and play continues with the remaining 44 counters.

Player Two rolls a 10 and divides the remaining 44 counters by 10. She has 10 groups of 4, with 4 counters leftover. Player Two keeps the 4 counters and returns the 40 remaining counters.

When no more divisions can be made, the game is over. The player with the most counters is the winner.

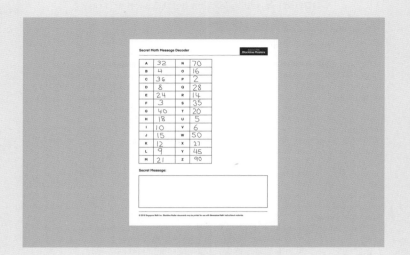

◀ **Exercise 9 • page 51**

Brain Works

★ Secret Math Messages

Materials: Secret Math Message Decoder (BLM)

Students fill in the decoder sheet with the products to math equations. They create a secret message and using the equations instead of letters, write the message next to the decoder.

Students trade decoder sheets with classmates and try to decode the message.

Example:

Secret message:

$\overline{\quad 3 \times 7 \quad}$ $\overline{\quad 4 \times 8 \quad}$ $\overline{\quad 4 \times 5 \quad}$ $\overline{\quad 6 \times 3 \quad}$

$\overline{\quad 7 \times 2 \quad}$ $\overline{\quad 4 \times 4 \quad}$ $\overline{\quad 4 \times 9 \quad}$ $\overline{\quad 6 \times 2 \quad}$ $\overline{\quad 5 \times 7 \quad}$!

Exercise 1 • pages 27–28

Chapter 9 Multiplication and Division of 3 and 4

Exercise 1

Basics

1. Count by threes and complete the multiplication equations.

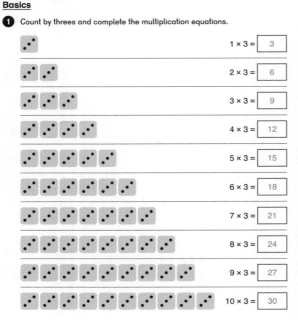

1 × 3 = 3
2 × 3 = 6
3 × 3 = 9
4 × 3 = 12
5 × 3 = 15
6 × 3 = 18
7 × 3 = 21
8 × 3 = 24
9 × 3 = 27
10 × 3 = 30

2. The sum of the digits in the products is __3__, __6__, or __9__.

Practice

3. (a) 3 × 3 is __3__ more than 2 × 3.
   3 × 3 = 9

   (b) 4 × 3 is 3 more than __3__ × 3.
   4 × 3 = 12

   (c) 5 × 3 = 15
   6 × 3 = 15 + 3 = 18
   7 × 3 = 15 + 6 = 21

   (d) 9 × 3 = 30 − 3 = 27
   8 × 3 = 30 − 6 = 24

4. Each cake has 3 candles. How many candles are on 7 cakes?

   7 × 3 = 21

   __21__ candles are on 7 cakes.

5. Circle products of 3.

   17   16   (12)   9   25   (21)   (18)

Exercise 2 • pages 29–30

Exercise 2

Basics

1.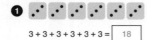

   3 + 3 + 3 + 3 + 3 + 3 = 18   6 + 6 + 6 = 18
   6 × 3 = 18   3 × 6 = 18

2. 8 × 3 = 24
   3 × 8 = 24

3. (a) 5 × 3 = 15 | 3 × 5 = 15
   (b) 2 × 3 = 6 | 3 × 2 = 6
   (c) 7 × 3 = 21 | 3 × 7 = 21
   (d) 10 × 3 = 30 | 3 × 10 = 30
   (e) 1 × 3 = 3 | 3 × 1 = 3
   (f) 9 × 3 = 27 | 3 × 9 = 27
   (g) 4 × 3 = 12 | 3 × 4 = 12
   (h) 3 × 3 = 9

Practice

4. Match.

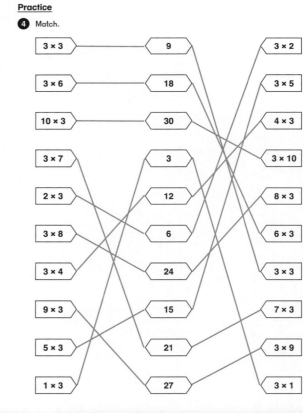

Exercise 3 • pages 31–34

Exercise 3

Basics

1 Dion is planting 3 tomato seeds in each jiffy pot.
How many jiffy pots does he need for 24 seeds?

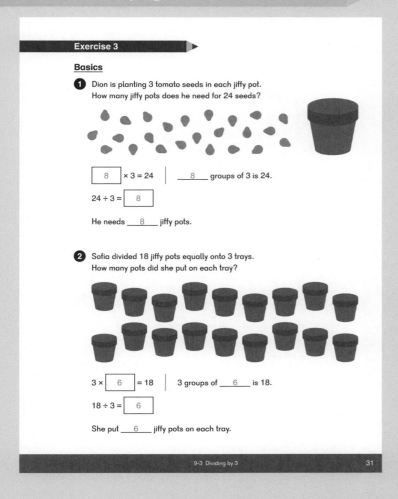

$\boxed{8} \times 3 = 24$ $\underline{\ 8\ }$ groups of 3 is 24.

$24 \div 3 = \boxed{8}$

He needs $\underline{\ 8\ }$ jiffy pots.

2 Sofia divided 18 jiffy pots equally onto 3 trays.
How many pots did she put on each tray?

$3 \times \boxed{6} = 18$ 3 groups of $\underline{\ 6\ }$ is 18.

$18 \div 3 = \boxed{6}$

She put $\underline{\ 6\ }$ jiffy pots on each tray.

Practice

3
$\boxed{4} \times 3 = 12$	$\boxed{3} \times 3 = 9$
$12 \div 3 = \boxed{4}$	$9 \div 3 = \boxed{3}$
$\boxed{10} \times 3 = 30$	$\boxed{5} \times 3 = 15$
$30 \div 3 = \boxed{10}$	$15 \div 3 = \boxed{5}$
$\boxed{8} \times 3 = 24$	$\boxed{1} \times 3 = 3$
$24 \div 3 = \boxed{8}$	$3 \div 3 = \boxed{1}$
$\boxed{6} \times 3 = 18$	$\boxed{2} \times 3 = 6$
$18 \div 3 = \boxed{6}$	$6 \div 3 = \boxed{2}$
$\boxed{7} \times 3 = 21$	$\boxed{9} \times 3 = 27$
$21 \div 3 = \boxed{7}$	$27 \div 3 = \boxed{9}$

4 (a) $\boxed{12} \div 3 = 4$ (b) $\boxed{27} \div 3 = 9$

(c) $\boxed{21} \div 3 = 7$ (d) $\boxed{15} \div 3 = 5$

(e) $\boxed{18} \div 3 = 6$ (f) $\boxed{24} \div 3 = 8$

5 Avery, Dana, and Grace share a box of 12 colored pencils equally.
How many pencils does each girl get?

$12 \div 3 = \boxed{4}$

Each girl gets $\underline{\ 4\ }$ pencils.

6 There are 21 tennis balls.
Sharif puts 3 tennis balls in each can.
How many cans does he need?

$\boxed{21} \div \boxed{3} = \boxed{7}$

He needs $\underline{\ 7\ }$ cans.

7 Laila has a ribbon that is 9 feet long.
She cuts it into 3 equal pieces.
How long is each piece?

$\boxed{9} \div \boxed{3} = \boxed{3}$

Each piece is $\underline{\ 3\ }$ feet long.

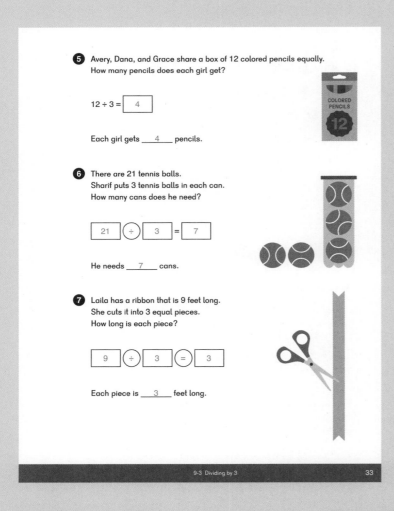

Challenge

8 Caleb has 2 boxes with 6 markers in each box.
He and two friends share them equally.
How many markers does each boy get?

$12 \div 3 = 4$

Each boy gets $\underline{\ 4\ }$ markers.

9 Mariya has 20 photos and 3 pages in her album.
If she wants to put the same number of photos on each page, what is the least number of photos she will have left over?

Students can use objects, draw pictures, or use equations to solve.

She will have $\underline{\ 2\ }$ photos left over.

10 Ryan is placing stakes 3 inches apart from each other.
The distance from the first to the last stake is 30 inches.
How many stakes has he placed?

Students can draw pictures, think logically, or use equations to solve. 30 inches divided by 3 is 10 spaces of 3 inches, but there is an extra stake at the start or end.

He has placed $\underline{\ 11\ }$ stakes.

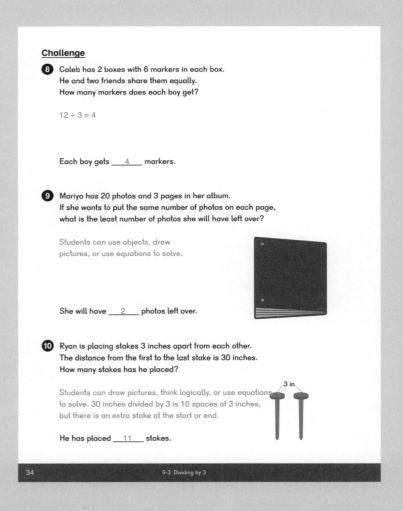

Exercise 4 • pages 35–38

Exercise 4

Check

1. Add or subtract.

   896 + 48 = 944   279 + 107 = 386   456 + 365 = 821

   148 − 82 = 66   432 − 63 = 369   803 − 29 = 774

2. Write +, −, ×, ÷, or = in each ◯.   Some answers may vary. For example, (a) could be 15 = 3 × 5.

   (a) 15 ÷ 3 = 5   (b) 15 = 3 + 12
   (c) 15 + 3 = 18   (d) 12 − 3 = 9
   (e) 24 − 3 = 21   (f) 21 ÷ 3 = 7
   (g) 7 + 3 = 10   (h) 10 × 3 = 30
   (i) 3 ÷ 3 = 1   (j) 3 + 3 = 6

3. Multiply or divide.

2 × 5 = 10	9 × 5 = 45	6 × 3 = 18
S	N	H
10 × 3 = 30	15 ÷ 3 = 5	27 ÷ 3 = 9
R	T	U
7 × 2 = 14	10 ÷ 5 = 2	3 × 2 = 6
R	E	S
4 × 3 = 12	3 × 7 = 21	18 ÷ 1 = 18
N	E	H
24 ÷ 3 = 8	9 ÷ 3 = 3	8 × 2 = 16
R	I	E
35 ÷ 5 = 7	5 × 10 = 50	20 ÷ 5 = 4
O	T	O

   Write the letters that match the answers above to learn a fun fact.

H	O	R	S	E	S		R	U	N		O	N
18	4	8	10	21	6		15	14	9		12	24	7	45

T	H	E	I	R		T	O	E	S
11	50	18	16	3		30	27	5	4	2	6	19	36

4. 3 bags of flour weigh 9 kilograms. How much does one bag of flour weigh?

   9 ÷ 3 = 3

   One bag of flour weighs __3__ kg.

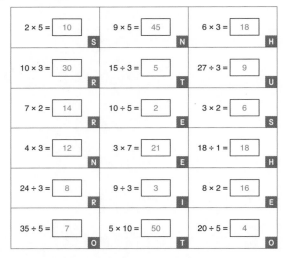

5. A bag of potatoes weighs 3 kilograms. How much do 9 bags of potatoes weigh?

   9 × 3 = 27

   9 bags of potatoes weigh __27__ kg.

6. Arman can fit 15 pots equally onto 3 trays.

   (a) How many pots go on each tray?

   15 ÷ 3 = 5

   __5__ pots go on each tray.

   (b) How many trays are needed for 25 pots?

   25 ÷ 5 = 5

   __5__ trays are needed for 25 pots.

Challenge

7. + + = 12   12 ÷ 3 = 4, so ◆ = 4

    + + + = 14   14 − 8 = 6, 6 ÷ 2 = 3, so ● = 3

    + + = 9

8. There are 10 tricycles and bicycles in all. If there are 23 wheels, how many are bicycles and how many are tricycles?

   Students can draw pictures or use equations to solve.
   If 2 wheels are put on each cycle, 20 wheels are used,
   leaving 3 wheels which would go on 3 of the cycles.

   There are __7__ bicycles and __3__ tricycles.

9. A piece of string is 20 ft long. It is cut into as many pieces as possible that are each 3 ft long. How many 3-ft long pieces are there? How long is the left over piece of string?

   Students can draw pictures or use equations to solve.
   18 is the greatest product for 3 that is less than 20.
   3 × 6 = 18
   20 − 18 = 2

   There are __6__ pieces that are 3 ft long.

   The left over piece of string is __2__ ft long.

Exercise 5 • pages 39–40

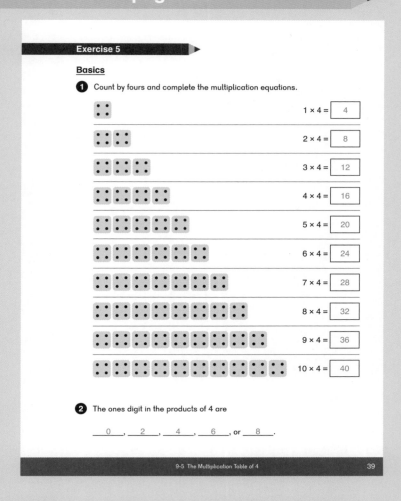

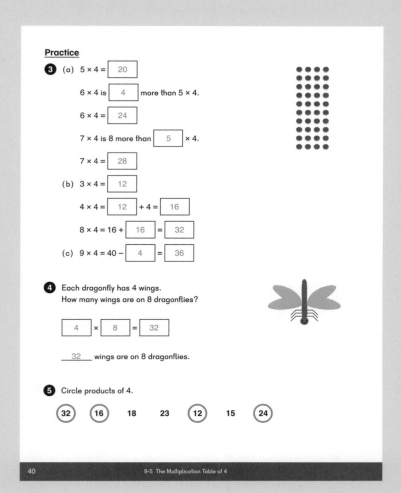

Exercise 6 • pages 41–42

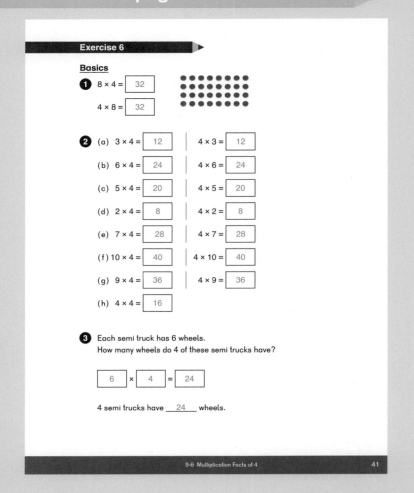

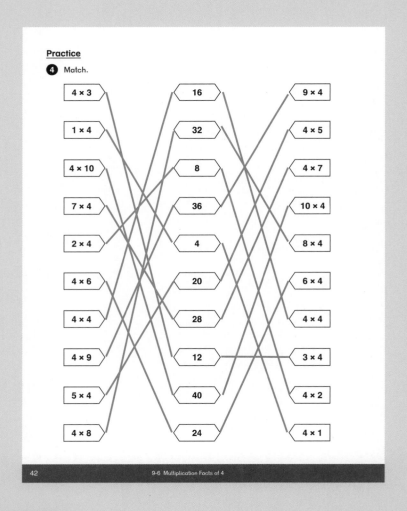

Exercise 7 • pages 43–46

Exercise 7

Basics

1. Alex is transplanting 32 tomato seedlings.

   (a) If he puts 4 seedlings in each row, how many rows will he have?

   $\boxed{8} \times 4 = 32 \quad | \quad 32 \div 4 = \boxed{8}$

   He will have __8__ rows.

   (b) If he wants to put them equally in 4 rows, how many seedlings should he put in each row?

   $4 \times \boxed{8} = 32 \quad | \quad 32 \div 4 = \boxed{8}$

   He should put __8__ seedlings in each row.

2. Mei is planting 4 melon seeds in each pot.

   (a) If she has 6 pots, how many melon seeds does she need?

   $\boxed{4} \times \boxed{6} = \boxed{24}$

   She needs __24__ seeds.

   (b) If she has 24 seeds, how many pots does she need?

   $\boxed{24} \div \boxed{4} = \boxed{6}$

   She needs __6__ pots.

Practice

3.
$\boxed{3} \times 4 = 12$	$\boxed{8} \times 4 = 32$
$12 \div 4 = \boxed{3}$	$32 \div 4 = \boxed{8}$
$\boxed{1} \times 4 = 4$	$\boxed{4} \times 4 = 16$
$4 \div 4 = \boxed{1}$	$16 \div 4 = \boxed{4}$
$\boxed{6} \times 4 = 24$	$\boxed{9} \times 4 = 36$
$24 \div 4 = \boxed{6}$	$36 \div 4 = \boxed{9}$
$\boxed{2} \times 4 = 8$	$\boxed{7} \times 4 = 28$
$8 \div 4 = \boxed{2}$	$28 \div 4 = \boxed{7}$
$\boxed{5} \times 4 = 20$	$\boxed{10} \times 4 = 40$
$20 \div 4 = \boxed{5}$	$40 \div 4 = \boxed{10}$

4. (a) $\boxed{16} \div 4 = 4$ (b) $\boxed{36} \div 4 = 9$

   (c) $\boxed{28} \div 4 = 7$ (d) $\boxed{20} \div 4 = 5$

   (e) $\boxed{24} \div 4 = 6$ (f) $\boxed{32} \div 4 = 8$

5. Jasper, Landon, Malik, and Wyatt share a box of 40 colored pencils equally. How many pencils does boy get?

   $40 \div \boxed{4} = \boxed{10}$

   Each boy gets __10__ pencils.

6. There are 16 basketballs. Grace puts them equally in 4 bags. How many basketballs are in each bag?

   $\boxed{16} \div \boxed{4} = \boxed{4}$

   __4__ basketballs are in each bag.

7. Mary has 28 pounds of flour. She puts 4 pounds of flour in each bag. How many bags does she need?

   $\boxed{28} \div \boxed{4} = \boxed{7}$

   She needs __7__ bags.

Challenge

8. (a) If $\blacklozenge \div 4 = 5$, what is $\blacklozenge \div 5$? $\boxed{4}$

   (b) If $\bullet \div \blacksquare = \bigstar$, what is $\bullet \div \bigstar$? $\boxed{\blacksquare}$

9. There are 10 ducks and goats in all. If there are 28 legs, how many ducks are there and how many goats are there?

   Each animal has 2 legs, using up 20 legs.
   There are still 8 legs. Put 2 each on 4 animals, so there are 4 goats.
   The rest are ducks with only 2 legs.

   There are __6__ ducks and __4__ goats.

10. Adriana has a tray filled with dirt 40 inches long and 12 inches wide. She wants to plant 1 seed every 4 inches along the length of the tray, and 4 inches from each edge of the tray. How many seeds can she plant?

    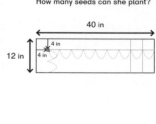

    Students may add seeds to the picture to solve. Students may divide 40 by 10, to get 10 lengths, so 9 seeds can be planted along one length. Similarly, there is room for one more rows. Students may also draw a grid and count intersections once they determine the number of horizontal and vertical lines needed.

    She can plant __18__ seeds.

Exercise 8 • pages 47–50

Exercise 8

Check

1 Add or subtract.

854 + 63 = 917   763 + 237 = 1,000   689 + 147 = 836

906 − 79 = 827   529 − 142 = 387   612 − 594 = 18

2 Write +, −, ×, ÷, or = in each ○.   Answers may vary.

(a) 20 ÷ 4 = 5   (b) 16 + 4 = 20
(c) 16 ÷ 4 = 4   (d) 12 − 4 = 8
(e) 12 = 4 × 3   (f) 28 + 4 = 32
(g) 28 = 4 × 7   (h) 8 × 4 = 32
(i) 8 − 4 = 4   (j) 8 ÷ 4 = 2

3 Multiply or divide.

3 × 4 = 12 **I**	4 ÷ 4 = 1 **E**	12 ÷ 4 = 3 **F**
12 ÷ 3 = 4 **R**	6 × 4 = 24 **T**	20 ÷ 4 = 5 **S**
4 × 8 = 32 **E**	24 ÷ 3 = 8 **K**	9 × 4 = 36 **C**
18 ÷ 2 = 9 **P**	28 ÷ 4 = 7 **D**	18 ÷ 3 = 6 **E**
4 × 4 = 16 **A**	8 ÷ 4 = 2 **W**	5 × 5 = 25 **H**
3 × 6 = 18 **O**	40 ÷ 4 = 10 **A**	45 ÷ 5 = 9 **P**
10 × 6 = 60 **E**	3 × 7 = 21 **H**	5 × 3 = 15 **A**

Joke: Why did the boy eat his math homework?
Write the letters that match the answers above to find out.

H	E		H	E	A	R	D		I	T
25	32		14	21	1	10	4		7	29	12	24

| W | A | S | | A | | P | I | E | C | E |
| 2 | 16 | 5 | | 26 | | 15 | 20 | 9 | 12 | 6 | 36 | 60 |

| | | O | F | | C | A | K | E | |
| 45 | 30 | 18 | 3 | | 27 | 36 | 16 | 8 | | 32 | 0 | 50 |

4 Ana bought 4 packs of pens.
There were 5 pens in each pack.
How many pens did she buy?

4 × 5 = 20

She bought ___20___ pens.

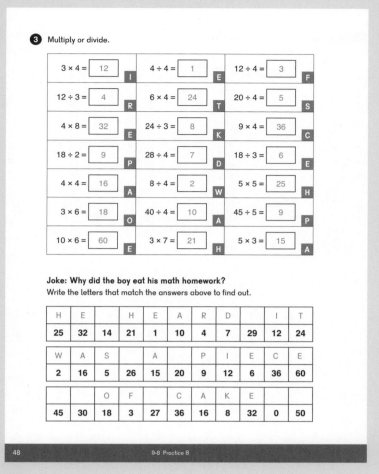

5 Dexter bought 4 packs of markers.
He bought 32 markers in all.
How many markers are in each pack?

32 ÷ 4 = 8

There are ___8___ markers in each pack.

6 Gary bought 4 pints of strawberries for $16.

(a) How much does 1 pint of strawberries cost?

16 ÷ 4 = 4

1 pint of strawberries costs $___4___.

(b) How much would 8 pints of strawberries cost?

8 × 4 = 32

8 pints of strawberries would cost $___32___.

Challenge

7 Mrs. Garcia made 34 empanadas.

(a) She wants to put them equally into 4 boxes.
How many empanadas will be left over?

Students can draw pictures, or think:
4 × 8 = 32, 4 × 9 = 36, so she can put 8 in
each box, which leaves 2.

___2___ empanadas will be left over.

(b) She gives away 2 full boxes of empanadas.
How many empanadas does she have left?

2 × 8 = 16
34 − 16 = 18

She has ___18___ empanadas left.

8 Clara has enough money to buy 4 comic books at $6 each.
Instead, she uses the same amount of money to buy some notebooks that cost $3 each.
How many notebooks does she buy?

4 × 6 = 24.
She has $24.
24 ÷ 3 = 8

She buys ___8___ notebooks.

Exercise 9 • pages 51–54

Exercise 9

Check

1 Find the values.

(a) 37 + 5 = **42**   (b) 659 + 7 = **666**

(c) 250 + 80 = **330**   (d) 598 + 50 = **648**

(e) 65 + 98 = **163**   (f) 99 + 52 = **151**

(g) 62 − 7 = **55**   (h) 691 − 8 = **683**

(i) 640 − 60 = **580**   (j) 351 − 80 = **271**

(k) 100 − 32 = **68**   (l) 303 − 97 = **206**

2 Write the missing numbers.

(a) **21** ÷ 3 = 7   (b) 24 ÷ **6** = 4

(c) 5 × **8** = 40   (d) **8** × 2 = 16

(e) 5 = 25 ÷ **5**   (f) 12 = 4 × **3**

---

3 Olga is planting corn, beans, and squash together.

(a) She is putting 5 corn seeds in each mound of dirt.
There are 8 mounds.
How many corn seeds does she use?

5 × 8 = 40
She uses 40 corn seeds.

(b) When the corn is 5 ft tall, she plants the same number of bean seeds around each mound.
She uses 32 bean seeds.
How many bean seeds does she plant around each mound?

32 ÷ 8 = 4
She plants 4 bean seeds around each mound.

(c) Then, she plants three yellow squash seeds and three green squash seeds around each mound.
How many of each type of squash seed is around all 8 mounds?

3 × 8 = 24
There are 24 yellow squash seeds and 24 green squash seeds in all.

(d) How many squash seeds did she plant in all?

24 + 24 = 48
She planted 48 squash seeds in all.

---

4 This table lists the cost of some bags of fruit.

Apples	$2
Grapes	$5
Bananas	$4
Oranges	$3
Lychee	$10

(a) How much do 6 bags of grapes cost?   **$30**

(b) How much do 6 bags of apples cost?   **$12**

(c) Melissa buys 6 bags of grapes and 6 bags of apples. How much does she spend?   **$42**

(d) How many bags of lychees can Travis buy with $40?   **4**

(e) How many bags of grapes can Travis buy with $40?   **8**

(f) How many more bags of grapes can Travis buy than bags of lychees?   **4**

(g) What is the total cost of 2 bags of each kind of fruit?   **$48**

2 + 5 + 4 + 3 + 10 = 24
24 + 24 = 48

Challenge

5 Complete the cross-number puzzle.

32	÷	4	=	8
−		+		+
20	÷	2	=	10
=		=		=
12	+	6	=	18

6 + + ✺ = 16   ● = 4

 + = 18   ◆ = 7

 = 8   ✺ = 5

45 ÷ ✺ = **9**

7 How many different products can be made by multiplying any two numbers below?

2 3
4 5

2 × 3 = 6
2 × 4 = 8
2 × 5 = 10
3 × 4 = 12
3 × 5 = 15
4 × 5 = 20
Six different products

Chapter 10 Money Overview

Suggested number of class periods: 8–9

	Lesson	Page	Resources	Objectives
	Chapter Opener	p. 69	TB: p. 47	Investigate dollars and cents.
1	Making $1	p. 70	TB: p. 48 WB: p. 55	Make $1.00 in different ways. Given an amount of money less than $1, mentally find the amount of money needed to make $1.
2	Dollars and Cents	p. 72	TB: p. 51 WB: p. 57	Count money up to $20 in bills and coins. Express money amounts using dollar notation ($), cent notation (¢), and dollar and cent notation (e.g., $1.25).
3	Making Change	p. 75	TB: p. 56 WB: p. 61	Make amounts of money within $20 in different ways. Convert from cents to dollars and cents or from dollars and cents to cents.
4	Comparing Money	p. 77	TB: p. 60 WB: p. 65	Compare amounts of money within $10.
5	Practice A	p. 79	TB: p. 64 WB: p. 69	Practice counting and comparing amounts of money.
6	Adding Money	p. 80	TB: p. 66 WB: p. 73	Add amounts of money within $10.
7	Subtracting Money	p. 83	TB: p. 70 WB: p. 77	Subtract amounts of money within $10.
8	Practice B	p. 86	TB: p. 74 WB: p. 81	Practice adding and subtracting money.
	Workbook Solutions	p. 88		

Chapter 10 Money

In **Dimensions Math® 1B** Chapter 18: Money, students learned:

- to identify coins and bills.
- the value of coins (penny, nickel, dime, and quarter).
- to count money amounts shown with coins within $1.
- the value of $10 and $5 bills.
- to count bills or coins up to $20.
- to add or subtract money in dollars or cents.

In this chapter, students will build on their knowledge by working with dollars and cents at the same time.

Lessons 6 and 7 cover adding and subtracting amounts of money within $20. While renaming the amounts to just cents and adding or subtracting with the vertical algorithm is always an option, the problems in these chapters are designed to emphasize and provide practice with the mental math strategies from Chapter 8.

Note that the term "decimal point" is not used here and should be avoided as students have not learned decimals yet. This chapter will use the informal term "dot." The dot separates the dollars from the cents.

Students will think of money as compound units: dollars and cents.

Students may either use paper strips or sticky notes, or draw the bar models if they need help figuring out word problems. They can also simply write an equation and solve. The bar models are provided as examples, not requirements.

Materials

- 100 pennies (counted out)
- 10-sided dice
- Bags
- Classroom items labeled with cost less than $10
- Newspapers and/or magazines
- Paper bag
- Price tags
- Real $1, $5, $10, and $20 bills
- Real pennies, nickels, dimes, and quarters
- Play coins and bills
- Recording sheets
- Whiteboards

Blackline Masters

- Number Bond Template
- What's a Word Worth?

Storybooks

- *Pigs Will Be Pigs: Fun with Math and Money* by Amy Axelrod
- *A Dollar, a Penny, How Much and How Many?* by Brian P. Cleary
- *Lemonade in Winter: A Book About Two Kids Counting Money* by Emily Jenkins
- *The Lunch Line* by Karen Berman Nagel
- *The Coin Counting Book* by Rozanne Lanczak Williams

Chapter Opener

Objective
- Investigate dollars and cents.

Lesson Materials
- Coins and bills
- Newspaper pages and/or magazines

Review the value of coins and bills with students.

Have students read and compare the costs of the crafts.

Provide students with newspapers or magazines and have them find prices of items.

Activity

▲ Trading Up

Materials: 10-sided die, play coins

Students start with one quarter. On her turn, a player rolls the dice and adds that amount of money to her start money.

If the amount added makes up another coin, she trades her change for the new coin.

The first player to get four quarters is the winner.

Sample play:

Player One rolls a 4. He adds 4 pennies to his quarter for 29 cents.

Player Two rolls a 6 and adds 6¢ to her money. She now has a quarter, a nickel, and a penny. (31¢)

Player One rolls a 6. He adds and exchanges his pennies. He now has a quarter and two nickels. He changes one nickel for a dime. (35¢)

Player Two rolls a 10. She collects pennies and makes exchanges to have a quarter, a dime, a nickel, and a penny. (41¢).

Lesson 1 Making $1

Objectives

- Make $1.00 in different ways.
- Given an amount of money less than $1, mentally find the amount of money needed to make $1.

Lesson Materials

- Play coins

Think

Provide students with play coins and pose the **Think** problem.

Have students share how they counted Mei and Sofia's coins, and how they found how much money each girl needs to have $1.00.

Learn

When counting money it is usually easiest to begin counting with the largest denomination.

Challenge students to follow Dion's suggestion to make $1 with coins in as many different ways as they can.

If additional practice is needed with counting coins, either provide students with different amounts of coins to count or play the game **Money Face-Off** shown on the following page.

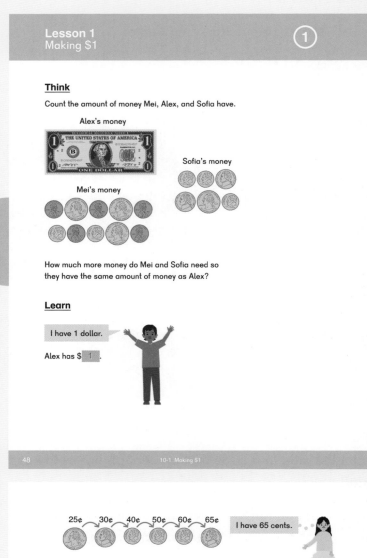

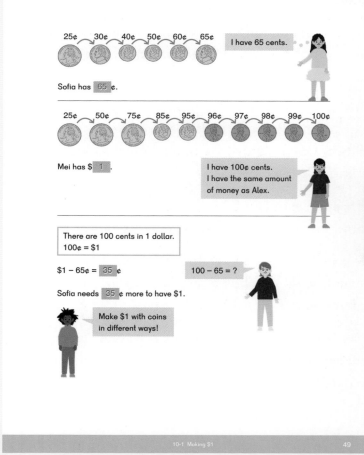

70   Teacher's Guide 2B Chapter 10   © 2017 Singapore Math Inc.

Do

Allow students to use play coins if needed. Most should be working from the textbook.

Activities

▲ **Money Face-Off**

Materials: Bag of random coins

Players take turns reaching into the bag and pulling out a handful of coins.

Players total the coins in each grab. The player with the highest value of coins scores a point.

The winner is the player with the most points after several rounds.

▲ **Stumper**

Materials: Play coins, Number Bond Template (BLM), $1 bill

Students play with partners. Using a Number Bond Template (BLM), students place the $1 bill in the "whole."

Partners take turns being the Stumper and the Solver.

The Stumper puts some coins in one part of the number bond. The Solver finds how much more is needed to make $1 and puts coins equaling that amount in the other part of the number bond.

If the Solver finds the correct value of coins for the missing part they get one point. If they make an error, the Stumper wins the point.

The winner is the player with the most points after the allotted time is up.

Exercise 1 • page 55

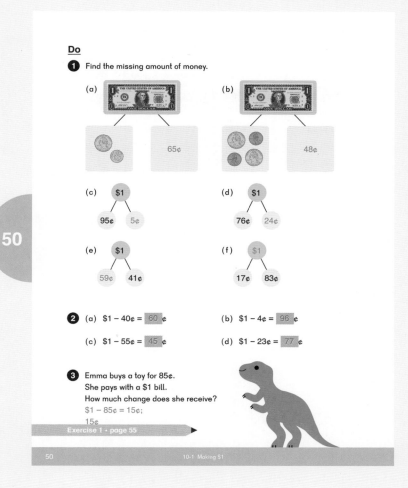

Lesson 2 Dollars and Cents

Objectives

- Count money up to $20 in bills and coins.
- Express money amounts using dollar notation ($), cent notation (¢), and dollar and cent notation (e.g., $1.25).

Lesson Materials

- Play coins and bills

Think

Pose the **Think** problem and provide students with play money if needed to work out how much Dion spent on the game.

Students should share the ways they counted the money.

Learn

Discuss the different ways that $17.53 is presented in numerals, words, and Emma's number bond.

Ensure students understand the ¢ and $ symbols and how they are used.

While the term "decimal point" is not introduced at this level, students may know the term. It is important that they know that the decimal point or dot is separating the dollars on the left from the cents on the right.

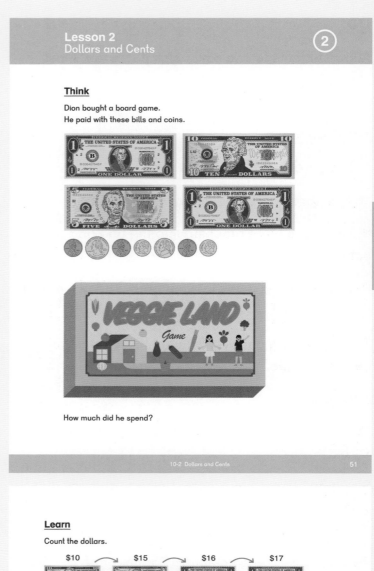

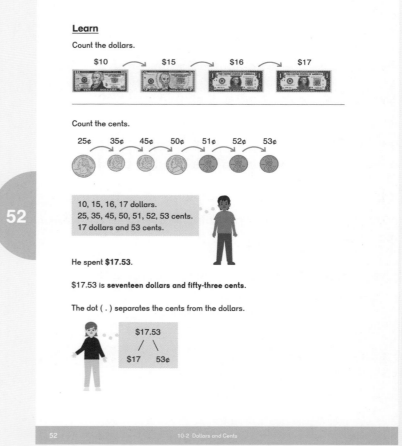

72   Teacher's Guide 2B Chapter 10   © 2017 Singapore Math Inc.

Do

When cents are written in dollars and cent notation, the maximum value is 99 cents. 100 cents is written as, "1 dollar" or, "$1.00."

❶ (f) When writing 45¢ in dollar and cents notation, students should see that there are zero dollars even though there is a dollar symbol. This is written as, "$0.45." We read this as, "45 cents" and not, "0 dollars and 45 cents."

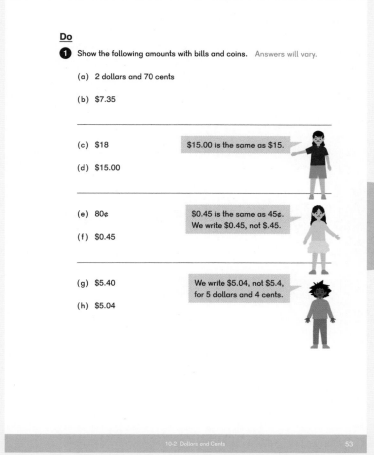

Exercise 2 • page 57

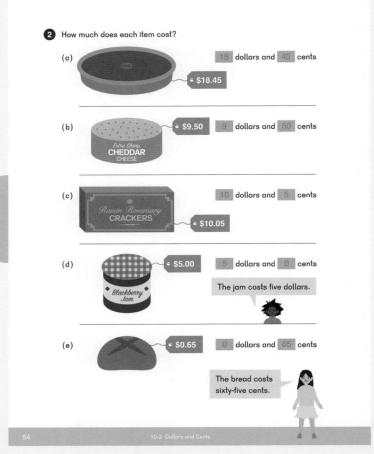

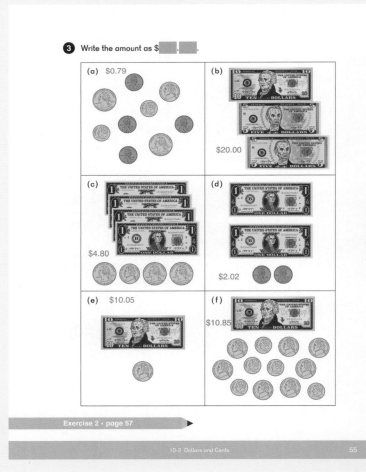

Lesson 3 Making Change

Objectives

- Make amounts of money within $20 in different ways.
- Convert from cents to dollars and cents or from dollars and cents to cents.

Lesson Materials

- Play coins and bills

Think

Provide students with play coins as needed to solve the **Think** problem. Discuss the ways students counted the money.

Learn

Emma used quarters to make 1 dollar, then added the rest of the change.

Ask students how many quarters there are in $2, $3, $4, and $5.

Do

Provide students play coins and bills and have them find different ways of making $5.

Have students discuss and count to verify the different ways they made $5 with coins and bills.

❶ Discuss the way that Emma made $5. Ask students why she can use 12 quarters in place of three $1 bills.

Mei reminds students that there are different ways of thinking about $5.00. Students can convert the dollars to all cents.

❷ Ask students how many cents Alex has (150¢).

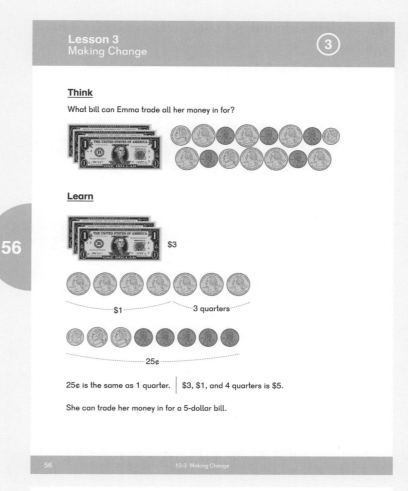

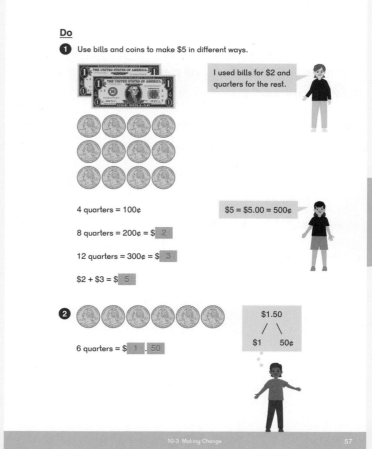

© 2017 Singapore Math Inc.   Teacher's Guide 2B Chapter 10

③–⑥ Students should not need play money for these problems.

⑦ (a) Guide students to look for patterns between dimes and pennies. Ask, "If 10 dimes is 100 cents, 20 dimes is…? And 30 dimes is…?"

Activity

▲ Dollar Nim

Materials: 100 pennies

Start with 100¢ in the "pot."

Two players take turns removing the value of a U.S. coin: penny, nickel, dime, or quarter.

The player who reduces the value of the pot to 0¢ is the winner.

Example play:

- Starting with 100 cents, Player One removes 25 cents worth of pennies. The pot is now worth 75 cents.
- Player Two removes 1 penny. The pot is now worth 74 cents.
- Player One removes 10 pennies, the pot is now worth 64 cents.

Play continues until the coins are gone. The player who takes the last remaining coin wins.

Exercise 3 • page 61

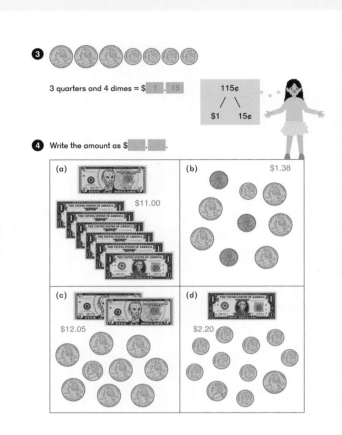

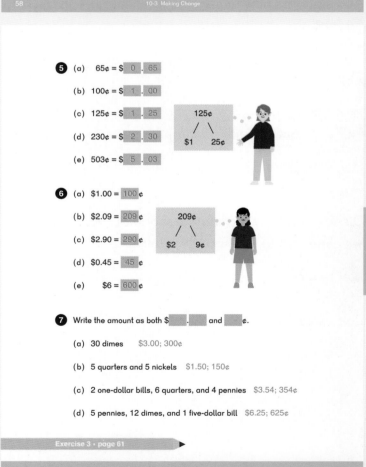

⑦ Write the amount as both $ ___ . ___ and ___ ¢.

(a) 30 dimes $3.00; 300¢

(b) 5 quarters and 5 nickels $1.50; 150¢

(c) 2 one-dollar bills, 6 quarters, and 4 pennies $3.54; 354¢

(d) 5 pennies, 12 dimes, and 1 five-dollar bill $6.25; 625¢

Lesson 4 Comparing Money

Objective

- Compare amounts of money within $10.

Lesson Materials

- Play coins and bills

Think

Pose the **Think** problems.

Ask students what strategies they have used to compare numbers that would help them compare amounts of money. Have students share their strategies.

Learn

When the dollar amounts are different, we can compare the dollars to see which amount is greater. Guide students to use comparison language:

- $7 is greater than $4.
- Emma has more money than Dion.
- Alex also has more money than Dion.

Students should realize that even though Alex has more coins and bills than Emma, he has less money than Emma. Remind them that the amount of money is not the same as the amount of coins.

Alex and Emma both have the same amount of dollars, so we need to look at the cents to compare.

- 26¢ is less than 27¢.
- Alex has less money than Emma.

Always read the sentence with the correct language:

- 26¢ < 27¢
  Twenty-six cents is less than twenty-seven cents.
- $7 > $4
  Seven dollars is greater (or more) than four dollars.

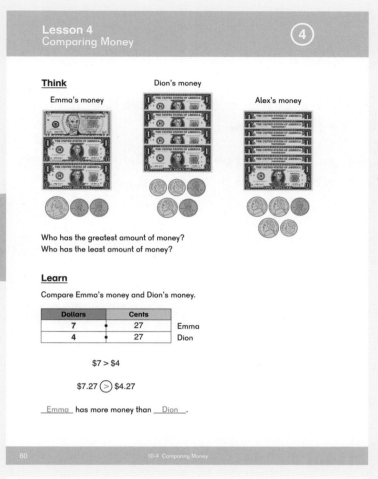

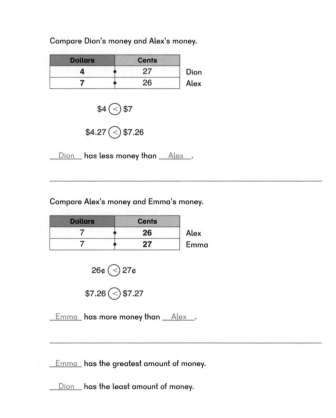

© 2017 Singapore Math Inc.   Teacher's Guide 2B Chapter 10   77

Do

❶ – ❷ Allow students who are struggling opportunities to use play money.

❺ (c – f) Ensure that students pay attention to the $ and ¢ signs in the problems.

❻ This question could be done as a center activity with the amounts written on index cards.

Exercise 4 • page 65

Do

❶ Which set has more money? B

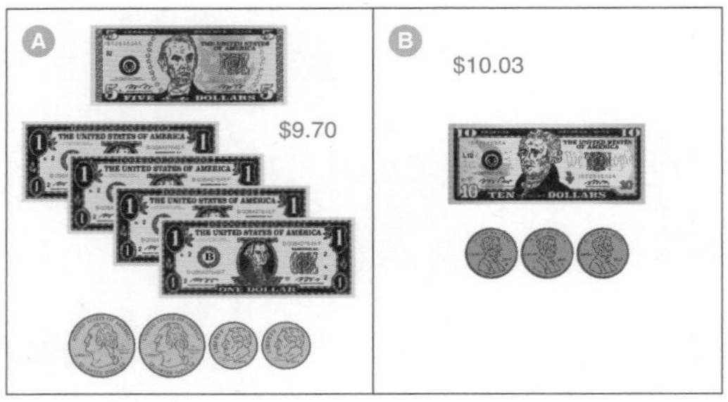

A $9.70
B $10.03

❷ Which set has less money? D

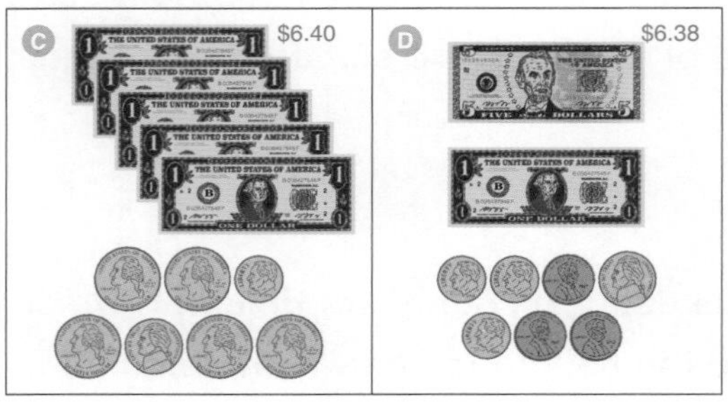

C $6.40
D $6.38

❸ Which roll of tape is less expensive? E

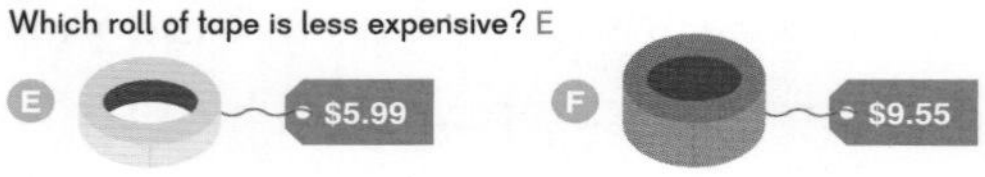

E $5.99
F $9.55

❹ Which dip is more expensive? G

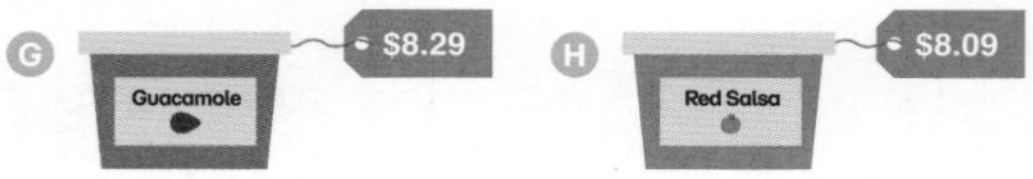

G Guacamole $8.29
H Red Salsa $8.09

❺ What sign, > or <, goes in the ◯?

(a) $8.99 < $9.30
(b) $5.68 < $5.90
(c) 75¢ > $0.69
(d) $4 > $0.40
(e) 103¢ < $1.30
(f) $5 > 63¢

❻ Put the price tags in order from least expensive to most expensive.

(a) $3.20, $2.99, $3.02, $9.29 → $2.99, $3.02, $3.20, $9.29
(b) $6.43, $6.34, $4.63, $3.46 → $3.46, $4.63, $6.34, $6.43
(c) $0.99, 63¢, $6, $6.03 → 63¢, $0.99, $6, $6.03

Exercise 4 • page 65

Lesson 5 Practice A

Objective

- Practice counting and comparing amounts of money.

After students complete the **Practice** in the textbook, have them continue to practice counting amounts of money with activities from the chapter.

In the next lessons, students will add and subtract dollars and cents.

Exercise 5 • page 69

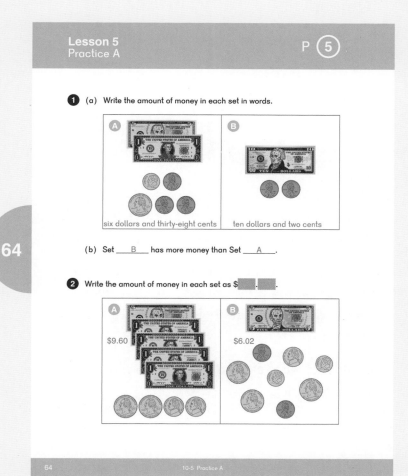

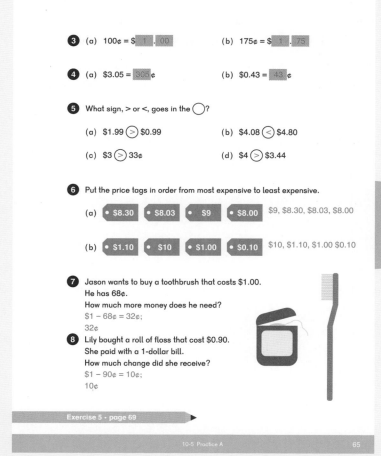

Lesson 6 Adding Money

Objective
- Add amounts of money within $10.

Lesson Materials
- Play coins and bills

Think

Provide students with play money. Have them figure out how much money Mei spent on the two items.

Observe how students are combining the amounts. Are they using the play money? Are they adding dollars to dollars and cents to cents? Are they counting on?

Ask students what strategies they used to add the amounts together.

Learn

Discuss the two methods with students.

Method 1 adds the dollars together, then adds the cents together, then combines the amounts. This can be a simpler strategy, especially when there are less than 100 cents and students do not need to regroup cents into a dollar.

Method 2 uses the greater priced item as a start and adds on, first the dollars, then the cents.

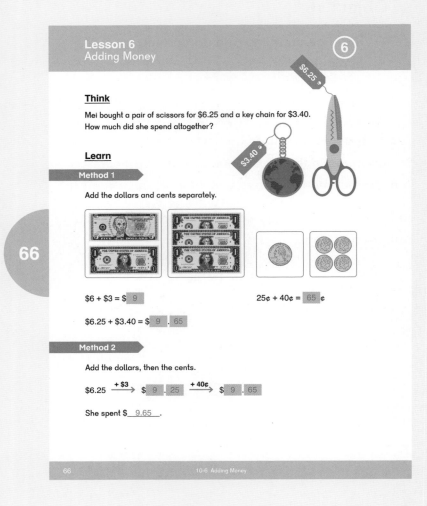

Do

Allow students who are struggling more opportunities to use play money.

Note that if using play money, many students will make the $0.35 with a quarter and a dime. If students need coins to work with, have them use only dollars, dimes, and pennies to relate the coins to the place value of hundreds, tens, and ones.

❷ Ensure that students pay attention to the $ and ¢ signs in the problems. Ask students if they used Method 1 or 2 from **Think** to find the total amount.

❸ These problems are preparing students for adding sums over 100 cents (❹) by first having them add exactly to 100 cents.

❹ Discuss Sofia and Alex's strategies with students.

Sofia's strategy is to make one hundred from the coins she has available.

Alex's strategy is to use number bonds and find what part of 55 makes a 100 with 65.

Note that it isn't always possible to make exactly one hundred with the given coins, but rearranging which coins are added can still simplify the sum.

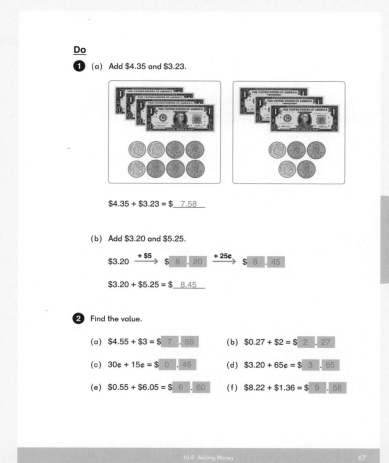

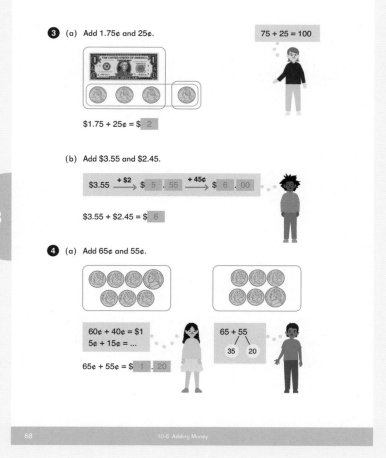

4 (b)
- Mei is adding whole dollars first: $4.85 + $2.
- Then she is adding cents to make the next dollar: $6.85 + 15¢.
- $7 + 35¢ = $7.35

A student may also add dollars to dollars and cents to cents:

- $4 + $2 = $6
- 85¢ + 50¢ = 135¢
- $6 + $1.35 = $7.35

5–**7** Have students share how they solved some of the problems.

In addition to the mental strategies, here are other methods that students may use based on their background knowledge of addition:

6 (f) 68 tens + 25 tens

(g)   155¢
    +455¢
    ─────
     610¢ = $6.10

(h)   750¢
    + 85¢
    ─────
     835¢ = $8.35

7 Students may want to draw a model to help them solve the problem.

Activity

▲ 20 Up

Materials: Play coins and bills (pennies, nickels, dimes, quarters, $1 bills, $5 bills) in a pile or paper bag, recording sheet for each player

Players each begin with $1 on their recording sheet. On the first turn, a player draws one bill and up to 5 coins. They find the total amount and add that number to their current amount. (So they are adding to the original $1.) The first player to go above $20 is the winner.

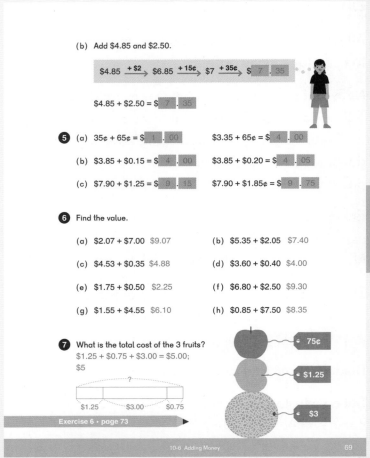

Lesson 7 Subtracting Money

Objective
- Subtract amounts of money within $10.

Lesson Materials
- Play coins and bills

Think

Provide students with play money and pose the **Think** problem.

Observe how students are subtracting the amounts. Ensure that they are subtracting dollars from dollars and cents from cents.

Ask students what strategies they used to find how much money Sofia has left.

Learn

Discuss the two methods with students. Similar to the previous lesson, students apply those strategies to subtraction.

Method 1 subtracts the dollars and cents separately. This can be a simpler strategy, especially when there is no regrouping of cents.

Method 2 subtracts first dollars, then cents.

Do

Give students who are struggling more opportunities to use coins and bills.

Note that if students are using coins, they might make the $0.65 with 2 quarters, a dime, and a nickel, which will make it harder to simply remove or "cross off" coins. For this reason, if students still need coins to work with, have them use only dollars, dimes, and pennies for hundreds, tens, and ones.

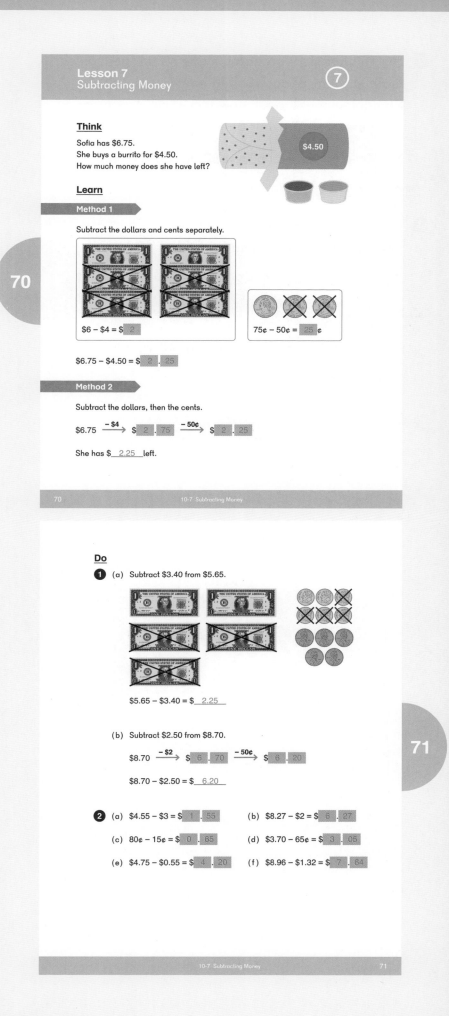

© 2017 Singapore Math Inc.   Teacher's Guide 2B Chapter 10   83

❸ (a) This problem prepares students for the mental math strategy of subtracting from one dollar. Students should recall how they have decomposed numbers from Chapter 8: Mental Math:

$3.00 − 40¢ = $2 + 60¢ = $2.60
    / \
$2 100¢

❹ (a) Just as students previously subtracted from 100, now Emma is subtracting from $1 or 100¢:

$4.20 − 80¢ = $3.20 + 20¢ = $3.40
   / \
$3.20 100¢

Students may also think of this problem as:

- 42 tens − 8 tens, or
- 42 tens − 10 tens + 2 tens, or
- $4.20 − $1.00 + $0.20.

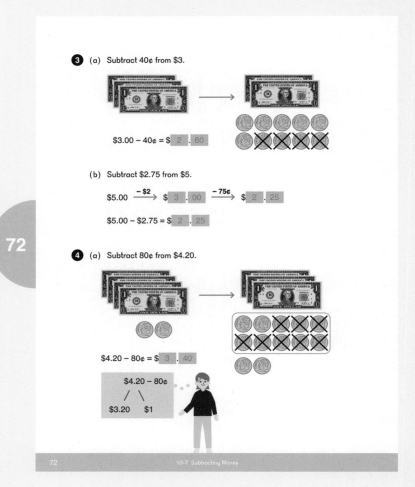

84　　　　　　　　　　Teacher's Guide 2B Chapter 10　　　　　　　　© 2017 Singapore Math Inc.

4 (b) Dion subtracts $2 to get $2.20, then splits the $2.20 to subtract 80¢ from $1.

5–**6** Have students share how they solved some of the problems.

Ensure that students pay attention to the $ and ¢ signs in the problems.

7 This is a 2-step problem. Suggest that students use a bar model if needed. They should see that they can either subtract the cost of the pear and peanuts from $7 separately, or add the cost of the pear and peanuts first, then subtract the total from $7.

Activity

▲ 20 Out

Materials: Play coins and bills (pennies, nickels, dimes, quarters, $1 bills, $5 bills) in a pile or paper bag, recording sheet

Players each begin with $20 on their recording sheet. On the first turn, a player draws one bill and up to 5 coins. They find the total amount drawn and subtract that number from their current amount. ($20 on first turn.) The first player to go below $1 is the winner.

Exercise 7 • page 77

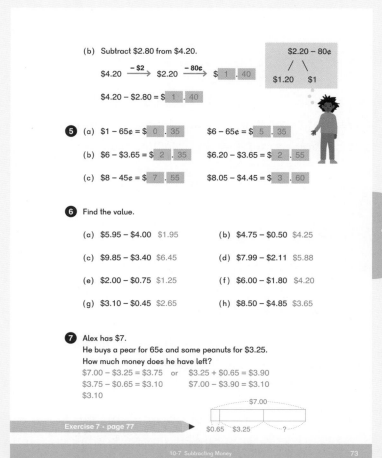

© 2017 Singapore Math Inc. Teacher's Guide 2B Chapter 10 85

Lesson 8 Practice B

Objective

- Practice adding and subtracting money.

After students complete **Practice** in the textbook, have them continue to practice mental addition and subtraction of money with activities from the chapter.

Activity

▲ **Shopping Spree**

Materials: Play coins and bills, items in classroom marked with price tags costing less than $10, recording sheet for each player

Pair students and provide a shopping list for each. Students walk around the room taking turns being the Buyer and the Seller. The Buyer starts with a $10 bill.

To make her purchase, the Buyer hands the Seller a $10 bill. The Seller subtracts the cost of the item from $10 and gives the Seller the difference in change. The Buyer records the equation on her recording sheet.

Each buyer continues to purchase items until they are out of money or do not have enough money left to buy anything.

Players switch roles.

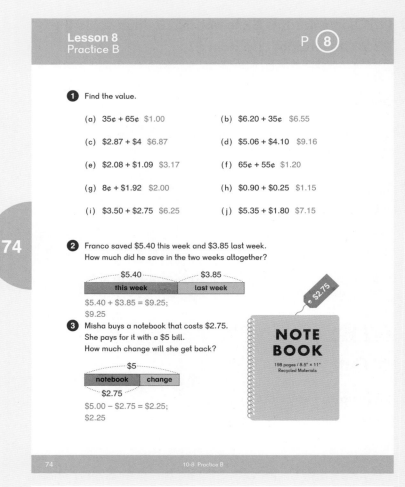

Exercise 8 • page 81

Brain Works

★ **What's a Word Worth?**

Materials: What's a Word Worth? (BLM)

Using What's a Word Worth? (BLM) as a guide, have students find the value of different words. They could use their names, spelling lists, or short phrases. Students can total up all the words on their spelling list each week and see what week is the most expensive.

Math rocks = 97¢

Sofia = 27¢

Mathematics = $1.01

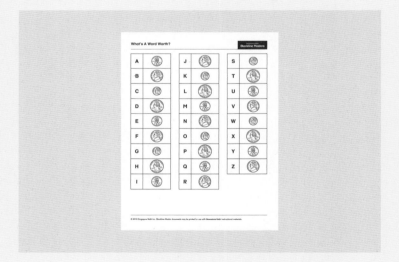

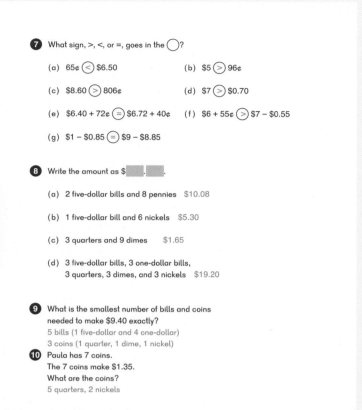

7 What sign, >, <, or =, goes in the ◯?

(a) 65¢ (<) $6.50 (b) $5 (>) 96¢
(c) $8.60 (>) 806¢ (d) $7 (>) $0.70
(e) $6.40 + 72¢ (=) $6.72 + 40¢ (f) $6 + 55¢ (>) $7 – $0.55
(g) $1 – $0.85 (=) $9 – $8.85

8 Write the amount as $▢.▢.

(a) 2 five-dollar bills and 8 pennies $10.08
(b) 1 five-dollar bill and 6 nickels $5.30
(c) 3 quarters and 9 dimes $1.65
(d) 3 five-dollar bills, 3 one-dollar bills, 3 quarters, 3 dimes, and 3 nickels $19.20

9 What is the smallest number of bills and coins needed to make $9.40 exactly?
5 bills (1 five-dollar and 4 one-dollar)
3 coins (1 quarter, 1 dime, 1 nickel)

10 Paula has 7 coins.
The 7 coins make $1.35.
What are the coins?
5 quarters, 2 nickels

Exercise 8 • page 81

Exercise 1 • pages 55–56

Chapter 10 Money

Exercise 1

Basics

1. 1 dollar = 100 cents

   $1 = [100] ¢

2. (a) Write the total cents for each set of coins.

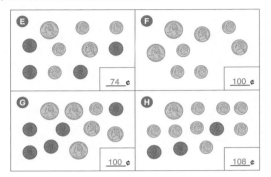

   E: 74 ¢
   F: 100 ¢
   G: 100 ¢
   H: 108 ¢

   (b) Which sets have the same amount of money as 1 dollar?
   F and G
   (c) Which set has more than 1 dollar?
   H
   (d) What coins can you add to the set that has less than 1 dollar to make $1?
   Answers will vary.
   Example: 1 quarter and 1 penny

Practice

3. Circle to separate the coins into 4 sets of $1.   Answers will vary.

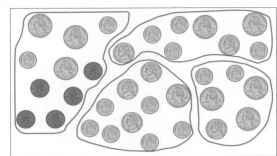

4. (a) 85¢ + [15] ¢ = 100¢   (b) 93¢ + [7] ¢ = $1
   (c) 24¢ + [76] ¢ = $1   (d) 7¢ + [93] ¢ = $1
   (e) $1 − 40¢ = [60] ¢   (f) $1 − 25¢ = [75] ¢
   (g) $1 − 53¢ = [47] ¢   (h) $1 − 22¢ = [78] ¢

5. Santino pays for a snack that cost 62¢ with $1.
   How much change does he get?

   $1 − 62¢ = 38¢
   He gets 38¢ in change.

Exercise 2 • pages 57–60

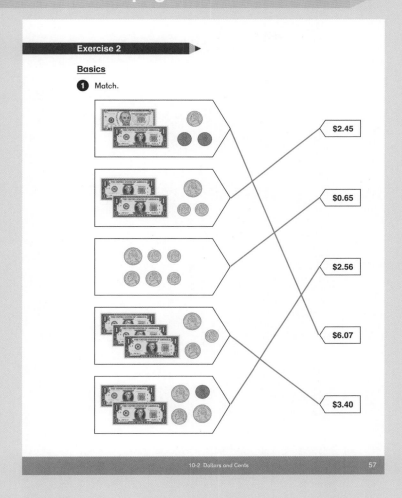

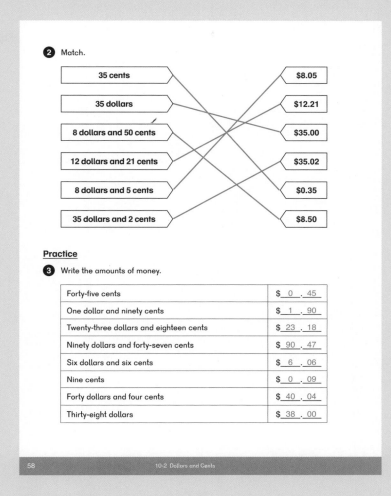

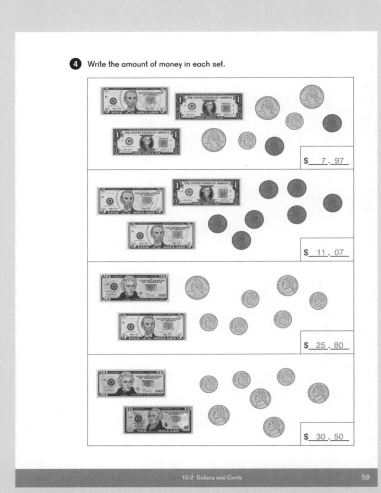

Exercise 3 • pages 61–64

Exercise 3

Basics

1. (a) 100¢ = $1.00   (b) 200¢ = $__2__.__00__

    25¢ = $0.25   5¢ = $__0__.__05__

    125¢ = $__1__.__25__   205¢ = $__2__.__05__

    (100¢) (25¢)   (200¢) (5¢)

2. Match.

   735¢ —— $6.05 (no, crosses)
   605¢ —— $7.35
   65¢ —— $6.50
   650¢ —— $0.05
   700¢ —— $7.65
   5¢ —— $0.65
   765¢ —— $7

   735¢ → $7.35
   605¢ → $6.05
   65¢ → $0.65
   650¢ → $6.50
   700¢ → $7
   5¢ → $0.05
   765¢ → $7.65

Practice

3. Write the amount of money.

   $ __1.31__

   $ __2.32__

   $ __4.46__

   $ __8.15__

4. Complete the tables.

Dollars	Cents
$4.32	432¢
$0.24	24¢
$0.08	8¢
$3.40	340¢
$8.02	802¢
$9.56	956¢
$6.20	620¢
$8	800¢

Cents	Dollars
231¢	$2.31
870¢	$8.70
708¢	$7.08
428¢	$4.28
75¢	$0.75
40¢	$0.40
400¢	$4.00
8¢	$0.08

5. (a) __8__ quarters make $2.00

   (b) __20__ dimes make $2.00

   (c) __40__ nickels make $2.00

   (d) __200__ pennies make $2.00

   (e) __11__ quarters make $2.75

   (f) __35__ dimes make $3.50

   (g) __32__ nickels make $1.60

6. Write the number of each type of bill or coin to make the given amount. Show three different ways for each amount.   Answers will vary. Examples given.

Amount	$5	$1	25¢	10¢	5¢
$1.30		1	1		1
			5		1
			4	3	
$2.75		2	3		
		1	7		
			8	7	1
$7.45	1	2	1	2	
	1	2		4	1
		7		4	1
$10	2				
		10			
	1	5			

Challenge

7. Kiara saves 1 quarter the first month, 2 quarters the second month, 3 quarters the third month, and so on. How much money will she save in 5 months?

   1 + 2 + 3 + 4 + 5 = 15
   15 quarters = $3.75

Exercise 4 • pages 65–68

Exercise 4

Basics

1 (a) Complete the tables and write >, <, or = in each ◯.

Amount	Dollars	Cents
$9.45	9	45
$7.62	7	62

$9 ⊖ $7 — >

$9.45 ⊖ $7.62 — >

Amount	Dollars	Cents
$3.45	3	45
$3.62	3	62

$3 ⊖ $3 — =

45¢ ⊖ 62¢ — <

$3.45 ⊖ $3.62 — <

(b) Arrange the amounts of money in order from least to greatest.

$9.45 $7.62 $3.45 $3.62

$3.45 $3.62 $7.62 $9.45

Practice

2 (a) Which set has more money? __A__

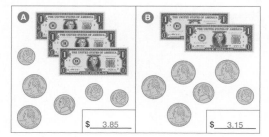

$ 3.85 $ 3.15

(b) Which set has less money? __D__

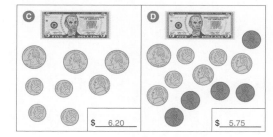

$ 6.20 $ 5.75

(c) Arrange the sets in order from greatest to least amount of money.

__B__ __A__ __D__ __C__

3 Circle the item that costs the most. Lunch box

$4.89 $8.25 $4.58

4 Circle the item that costs the least. Doll

$21.42 $12.42 $12.24

5 Write >, <, or = in each ◯.

(a) $5.62 ⊖ $6.94 — <
(b) $6.81 ⊖ $6.48 — >
(c) $7.03 ⊖ $3.07 — >
(d) $60.20 ⊖ $6.65 — >
(e) 854¢ ⊖ 812¢ — >
(f) 123¢ ⊖ $123 — <
(g) $1.02 ⊖ 102¢ — =
(h) 8¢ ⊖ $8 — <
(i) $0.71 ⊖ 83¢ — <
(j) $4 ⊖ 349¢ — >

6 The table shows how much money 5 people saved. Arrange the amounts of money in order. Start with the greatest amount.

Kalama	$6.23
Nolan	$5.99
Madison	$8.34
Pablo	$6.32
Jody	$9.34

$9.34 $8.34 $6.32 $6.23 $5.99

7 Circle the amounts of money that are greater than $2.45 but less than $6.

$2.35   ($2.85)   ($5.62) $6.07   ($3.86) $7.42

Challenge

8 Riya has 1 five-dollar bill, 2 one-dollar bills, 5 quarters, and 12 dimes.
Nora has 5 one-dollar bills, 12 quarters, and 5 dimes.
Who has more money?

Riya has $9.45.
Nora has $8.50
$9.45 > $8.50
Riya has more money.

Exercise 5 • pages 69–72

Exercise 5

Check

1 Practice addition and subtraction.

198 + 49 = 247   327 + 327 = 654   483 + 217 = 700

304 – 38 = 266   932 – 762 = 170   425 – 89 = 336

2 Color coins in each set to make $1 two different ways. Answers will vary.

3 (a) Write the amount of money in each set.

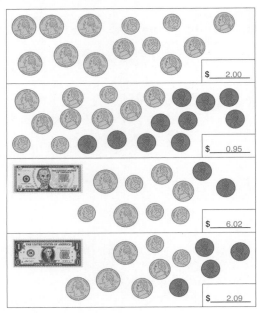

$ 2.00
$ 0.95
$ 6.02
$ 2.09

(b) Arrange the amounts of money in order from least to greatest.

$0.95 $2.00 $2.09 $6.02

4 Write the number of each type of bill or coin to make the given amount. Show four different ways for each amount. Answers will vary. Examples given.

Amount	$5	$1	25¢	10¢	5¢
$1.80		1	3		1
		1		8	
			6	3	
				18	
$3.35		3	3	3	
		3		3	5
		3	5	1	
		2		13	5
$5	1				
		5			
			4	1	
				50	
$8.50	1	3	2		
		8		5	
		7	4		10
		6	8	5	

5 Write >, <, or = in each ◯.

(a) $5.62 < $6.94   (b) $0.10 > 1¢

(c) $3.04 < $4.90   (d) $26 > 260¢

(e) 824¢ < $842   (f) 430¢ = $4.30

Challenge

6 Natalia bought a toy for $0.85. She paid with a 5-dollar bill. She got 3 bills and 8 coins in change. What coins were they?

$5 – $0.85 = $4.15
The 3 bills would be $3, so $1.15 in coins.
3 quarters, 3 dimes, 2 nickels

7 Find 2 ways to make $1 using 19 coins (quarters, dimes, nickels, or pennies) exactly.

Possible answers:
1 dime, 18 nickels
1 quarter, 1 dime, 12 nickels, 5 pennies
5 dimes, 9 nickels, 5 pennies
2 quarters, 1 dime, 6 nickels, 10 pennies
1 quarter, 5 dimes, 3 nickels, 10 pennies
9 dimes, 10 pennies
3 quarters, 1 dime, 15 pennies

Exercise 6 • pages 73–76

Exercise 6

Basics

1 (a) Add $5.70 and $2.25.

$5.70 →(+$2)→ $ 7.70 →(+25¢)→ $ 7.95

(b) Add $5.70 and $2.30.

$5.70 →(+$2)→ $ 7.70 →(+30¢)→ $ 8.00

(c) Add $5.70 and $2.45.

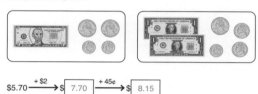

$5.70 →(+$2)→ $ 7.70 →(+45¢)→ $ 8.15

2 65¢ + 35¢ = $ 1.00 $4.65 + $0.35 = $ 5.00

65¢ + 50¢ = $ 1.15 $4.65 + $0.50 = $ 5.15

(35¢) (15¢)   (35¢) (15¢)

$4.65 + $3 = $ 7.65 $4.65 + $3.50 = $ 8.15

Practice

3 (a) $0.85 + $3 = $ 3.85   (b) $3.45 + $2 = $ 5.45

(c) $6.75 + 15¢ = $ 6.90   (d) $2.45 + 55¢ = $ 3.00

(e) $1.60 + 50¢ = $ 2.10   (f) $4.85 + 40¢ = $ 5.25

4 Write the missing numbers.

(a) $2.35 →(+$4)→ $ 6.35 →(+65¢)→ 7.00

$2.35 + $4.65 = 7.00

(b) $7.20 →(+$2)→ $ 9.20 →(+95¢)→ 10.15

$7.20 + $2.95 = 10.15

5 Add.

30¢ + 45¢ = 75¢	$1.22 + 20¢ = $1.42	$2.50 + $1.40 = $3.90
$8.60 + $0.04 = $8.64	$4.11 + $1.50 = $5.61	$5.55 + $2.80 = $8.35
75¢ + $3.51 = $4.26	$7.10 + $2.90 = $10.00	$1.89 + $0.75 = $2.64
$1.95 + $1.95 = $3.90	$1.49 + $0.99 = $2.48	$9.19 + 14¢ = $9.33
$3.40 + $2.25 = $5.65	$0.95 + $2.60 = $3.55	$5.10 + 45¢ = $5.55

How many hearts does an octopus have?
Color the spaces that contain the answers to find out.

$2.20	$6.00	$2.48	$8.35	$9.33	75¢	$5.23	99¢
$9.60	$3.33	$9.99	$7.80	$1.25	$4.26	$4.76	$2.75
$2.59	19¢	$8.90	$5.98	22¢	$5.65	$7.00	$9.50
$7.25	$4.02	50¢	$1.42	$3.90	$5.55	$2.20	$1.85
$6.30	$2.45	$2.40	$3.77	$9.20	$3.90	54¢	$9.50
$8.10	57¢	$8.15	$6.12	$7.65	$10.00	$4.19	$3.89
$5.27	$1.99	$5.61	$8.64	$2.64	$3.55	$6.15	$7.45

6 Aisha bought the cinnamon roll and the pack of gum.
How much did she spend?

$3.20 + $0.85 = $4.05
She spent $4.05.

7 Carter bought the muffin and the pretzel.
How much did he spend?

$1.80 + $3.75 = $5.55
Carter spent $5.55.

8 Dexter bought 2 of the items above and paid with a five-dollar bill.
He did not get any change.
Which two items did he buy?

$1.80 + $3.20 = $5
Dexter bought the muffin and the cinnamon roll.

9 Valentina bought all 4 items above.
How much did she spend?

$1.80 + $3.20 + $3.75 + $0.85 = $9.60
She spent $9.60.

Exercise 7 • pages 77–80

Exercise 7

Basics

1. (a) Subtract $1.20 from $5.50.

$5.50 →⁻$¹ $ 4.50 →⁻²⁰¢ $ 4.30

(b) Subtract $1.50 from $5.50.

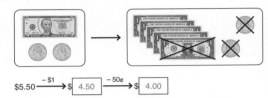

$5.50 →⁻$¹ $ 4.50 →⁻⁵⁰¢ $ 4.00

(c) Subtract $1.75 from $5.50.

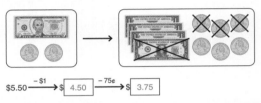

$5.50 →⁻$¹ $ 4.50 →⁻⁷⁵¢ $ 3.75

2. $6.00 − 35¢ = $ 5.65 $6.20 − 35¢ = $ 5.85

$6.20 − $3 = $ 3.20 $6.20 − $3.35 = $ 2.85

Practice

3. (a) $1.00 − 45¢ = 55 ¢   (b) $1.00 − $0.75 = $ 0.25
   (c) $5.00 − 15¢ = $ 4.85   (d) $7.00 − $0.55 = $ 6.45
   (e) $1.10 − 25¢ = $ 0.85   (f) $4.15 − $0.40 = $ 3.75

4. Write the missing numbers.

   (a) $7.00 →⁻$⁴ $ 3.00 →⁻⁶⁵¢ $ 2.35
       $7.00 − $4.65 = $ 2.35

   (b) $7.95 →⁻$² $ 9.95 →⁻²⁰¢ $ 9.75
       $7.95 − $2.20 = $ 9.75

   (c) $7.20 →⁻$² $ 5.20 →⁻⁹⁵¢ $ 4.25
       $7.20 − $2.95 = $ 4.25

5. Subtract.

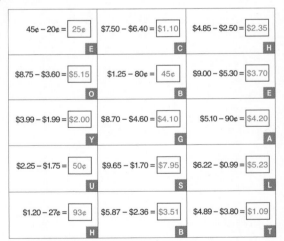

45¢ − 20¢ = 25¢	$7.50 − $6.40 = $1.10	$4.85 − $2.50 = $2.35
E	C	H
$8.75 − $3.60 = $5.15	$1.25 − 80¢ = 45¢	$9.00 − $5.30 = $3.70
O	B	E
$3.99 − $1.99 = $2.00	$8.70 − $4.60 = $4.10	$5.10 − 90¢ = $4.20
Y	G	A
$2.25 − $1.75 = 50¢	$9.65 − $1.70 = $7.95	$6.22 − $0.99 = $5.23
U	S	L
$1.20 − 27¢ = 93¢	$5.87 − $2.36 = $3.51	$4.89 − $3.80 = $1.09
H	B	T

Joke: Why isn't it fun to play basketball with pigs?
Write the letters that match the answers above to find out.

B	E	C	A	U	S	E		T	H	E	Y
45¢	25¢	$1.10	$4.20	50¢	$7.95	$3.70		$5.22	$1.09	$2.35	$3.70	$2.00

H	O	G		T	H	E		B	A	L	L
93¢	$5.15	$4.10	$1.01	$1.09	$2.35	25¢	99¢	$3.51	$4.20	$5.23	$5.23

6. Austin bought the soft pretzel.
   He paid with a five-dollar bill.
   How much change did he receive?

   $5.00 − $3.75 = $1.25
   He received $1.25 change.

7. How much more is the cinnamon roll than the pack of gum?

   $3.20 − $0.85 = $2.35
   It is $2.35 more.

Challenge

8. Papina bought 2 items above and paid with a five-dollar bill.
   She got 40¢ in change.
   Which two items did she buy?

   $5.00 − $0.40 = $4.60
   $3.75 + $0.85 = $4.60
   She bought the soft pretzel and the pack of gum.

Teacher's Guide 2B Chapter 10

Exercise 8 • pages 81–84

Exercise 8

Check

1. Color bills and coins to make the amounts of money.

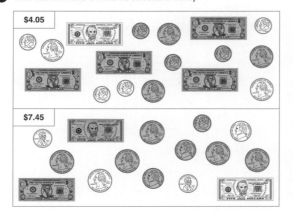

2. Write the amounts of money.

12 dimes and 3 nickels	$ 1.35
2 one-dollar bills and 5 quarters	$ 3.25
1 five-dollar bill, 2 one-dollar bills, and 3 quarters	$ 7.75
7 quarters and 7 dimes	$ 2.45
25 nickels and 25 pennies	$ 1.50

3. The answer to each problem is the first number in the next problem. Go through all the calculations in order to reach the goal of $10.00.

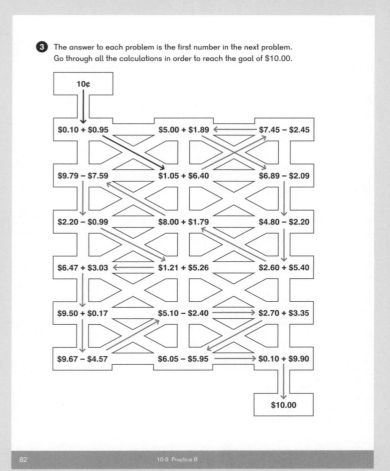

4. A toy car costs $6.20.
A toy robot costs $1.55 more than the toy car.
How much does the toy robot cost?

$6.20 + $1.55 = $7.75
The toy robot costs $7.75.

5. Andrei has $6.10.
He has 80¢ more than Daren.
How much money does Daren have?

$6.10 − 80¢ = $5.30
Daren has $5.30.

Challenge

6. Nina bought a hair bow for $0.80, a bracelet for $0.75, and a purse for $2.
She had $4.30 left.
How much money did she have at first?

$0.80 + $0.75 + $2 = $3.55 (how much she spent)
$3.55 + $4.30 = $7.85 (how much she had at first)
Nina had $7.85 at first.

Challenge

7. Circle to separate the bills and coins into 4 sets of $3.00. Answers will vary.

Notes

Chapter 11 Fractions — Overview

Suggested number of class periods: 7–8

	Lesson	Page	Resources		Objectives
	Chapter Opener	p. 101	TB:	p. 77	Investigate fractions.
1	Halves and Fourths	p. 102	TB: WB:	p. 78 p. 85	Identify halves and quarters as 2 or 4 equally divided parts of a whole. Recognize halves and quarters of shapes. Write halves and quarters using fractional notation.
2	Writing Unit Fractions	p. 105	TB: WB:	p. 81 p. 87	Recognize and name unit fractions up to $\frac{1}{10}$.
3	Writing Fractions	p. 107	TB: WB:	p. 84 p. 91	Write common fractions up to tenths.
4	Fractions that Make 1 Whole	p. 109	TB: WB:	p. 87 p. 95	Find two fractions that make 1 whole.
5	Comparing and Ordering Fractions	p. 111	TB: WB:	p. 89 p. 97	Compare and order fractions to tenths.
6	Practice	p. 113	TB: WB:	p. 92 p. 101	Practice comparing and ordering fractions. Practice finding fractions that make 1 whole.
	Review 3	p. 115	TB: WB:	p. 94 p. 105	Cumulative review of content from Chapter 1 through Chapter 11.
	Workbook Solutions	p. 117			

Chapter 11 Fractions

In **Dimensions Math® 1B** Chapter 15: Fractions, students learned to identify and divide shapes into halves and fourths.

In this chapter, students will learn fractional notation, however, the terms "numerator" and "denominator" will not be introduced until **Dimensions Math® 3B**. This is so the terminology does not interfere with understanding.

Fractional notation tells us how much of the whole a part represents. The denominator names the fractional unit (e.g., fourths) and the numerator counts the number of fractional units (e.g., 3 fourths).

Unit fractions represent one part of the whole. $\frac{1}{3}$ and $\frac{1}{7}$ are examples of unit fractions. Students will not need to know the term "unit fractions." They should, however, understand that a common fraction is composed of unit fractions: $\frac{3}{4} = \frac{1}{4} + \frac{1}{4} + \frac{1}{4}$.

Students will learn to compare unit fractions. To compare fractions, the whole for both fractions must be the same. It is important for students to realize that the unit fraction with the greater denominator is the smaller fraction, since the whole is divided into more parts. $\frac{1}{7}$ is smaller, or less, than $\frac{1}{3}$ because the whole is divided into 7 parts, which is more than 3 parts.

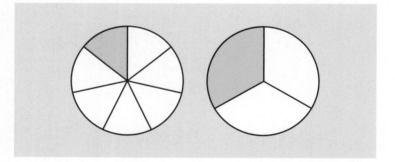

Previously, students used the word "whole" to refer to the total amount. For example, two parts make the whole in addition and subtraction, or equal parts make the whole in multiplication and division.

In the **Dimensions Math®** series, students first learn about fractions using shapes or objects, and the word "whole" refers to the entire shape or object. In later grades, the whole for a fraction will generally be the more abstract number 1, $\frac{1}{4}$ is $\frac{1}{4}$ of 1. Later, as students start calculating with fractions, the term "whole" may again refer to a quantity, i.e., they will be finding $\frac{1}{4}$ of 24 (multiplication of a fraction by a number).

Chapter 11 Fractions

Materials

- Colored pencils
- Construction paper
- Crayons or markers
- Fraction tiles
- Fraction circles
- Geometric shapes
- Googly eyes
- Paper shapes
- Pattern blocks
- Round paper plates or papers
- Scissors
- Square paper, such as Origami paper
- Square pieces of blank paper
- Strips of paper (strips handed out in individual lessons should be of equal length)
- Whiteboards

Blackline Masters

- Fraction Exercises
- Dot Paper
- Quilt Square
- Fraction Match Cards
- Triangle Graph Paper

Storybooks

- *A Fraction's Goal — Parts of a Whole* by Brian P. Cleary
- *Full House* by Dayle Ann Dodds
- *One Big Pair of Underwear* by Laura Gehl and Tom Lichtenheld
- *Mary Clare Likes to Share* by Joy Hulme
- *Jump, Kangaroo, Jump!* by Stuart J. Murphy
- *The Missing Piece* by Shel Silverstein
- *The Wishing Club* by Donna Jo Napoli and Anna Currey

Notes

Chapter Opener

Objective

- Investigate fractions.

Lesson Materials

- Paper shapes including squares, rectangles, circles, ovals, and odd shapes

Discuss the **Chapter Opener**. Ask students what it means to divide something equally.

Ask students, "What if there were more than 4 children?"

- Would they cut the cake differently?
- Would each child get the same size piece as if there had been only 4 children?

Provide students with a variety of paper shapes to represent differently shaped cakes.

Ask students to discuss which shapes would be easy to divide into four equal pieces, and which would be harder. Ask them to explain their answers.

Ask students to share what they have learned or may recall about fractions.

Lesson 1 Halves and Fourths

Objectives

- Identify halves and quarters as 2 or 4 equally divided parts of a whole.
- Recognize halves and quarters of shapes.
- Write halves and quarters using fractional notation.

Lesson Materials

- Sheets of square paper, such as Origami paper, several per student
- Paper circles or small paper plates
- Scissors
- Fraction Exercises (BLM)

Think

Have students discuss Alex and Mei's questions in **Think**.

Provide students with square pieces of paper and scissors. Ask them to cut the paper into 2 equal parts. See whether students simply cut, or first fold the paper and then cut it. Ask those that folded the paper first why they folded it. They may answer that by folding it carefully, they can see the two parts are equal.

Compare students' results. They may have cut the paper in different ways. Ask them to prove that the two parts are equal (by laying them on top of each other).

If all students folded to make 2 equal pieces, provide examples where the two pieces are not quite equal.

Repeat with fourths. Some students might first cut halves, and then put each half on top of the other to cut again into fourths. Others might fold twice. Have students save their parts (halves and wholes).

Pose the two questions in **Think**:

- How big is each part compared to the whole? (Each part is one-half or one-fourth of a whole.)
- How can we write the answer as a number?

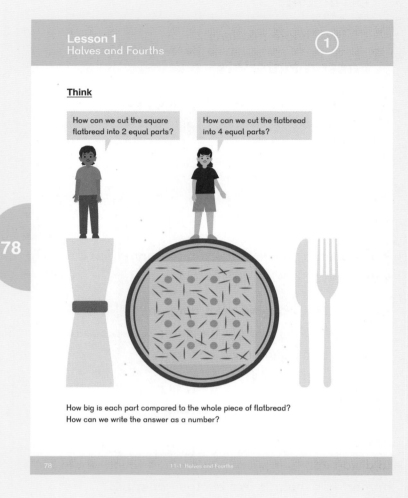

Discuss their answers. Some students may already know how to write fractions.

Learn

Tell students that "one-half" and "one-fourth" are fractions. One-half means 1 out of 2 equal parts, and one-fourth means 1 out of 4 equal parts.

Sofia points out that "fractions are numbers." Students may be confused now that there are two numerals and a symbol for a number. Tell students a fraction is still a number, the way 23 is a number with two digits.

Have students compare the papers they cut up in **Think** to the illustrations of paper in **Learn**.

Discuss the way fractions are written:

- $\frac{1}{2}$ means 1 out of 2 equally divided parts of the whole.
- $\frac{1}{4}$ is 1 out of 4 equally divided parts of the whole.

Have students label their parts from **Think** with the fraction and place them back together to make the whole.

Students should lay one part of each square on top of each other to see that $\frac{1}{2}$ is greater than $\frac{1}{4}$.

Ask students, "How many parts make a whole when a paper was divided into halves?" (2 parts of $\frac{1}{2}$ each make 1 whole.)

Repeat with fourths.

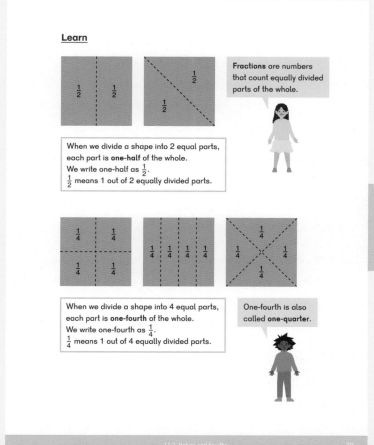

Do

1 Provide students with round paper plates or circles, all the same size. After students complete (a) through (d), they should lay one part of a quarter-circle on top of a half-circle to see that $\frac{1}{2}$ is greater than $\frac{1}{4}$.

Point out to students that in comparing $\frac{1}{2}$ to $\frac{1}{4}$, the circles that made up the whole were the same size. Remind them that they can only compare fractions from the same size whole.

Students can imagine a scenario where a cake as big as a table was cut into fourths, and another cake the size of a plate was cut into halves. $\frac{1}{4}$ of the large cake would be larger than $\frac{1}{2}$ of the small cake.

2 — **3** Students who struggle can work with Fraction Exercises (BLM) which has enlarged images of these shapes. Students can cut the parts out and compare. For example, some students may not recognize that **2** (c) shows halves without physically rotating the pieces to compare.

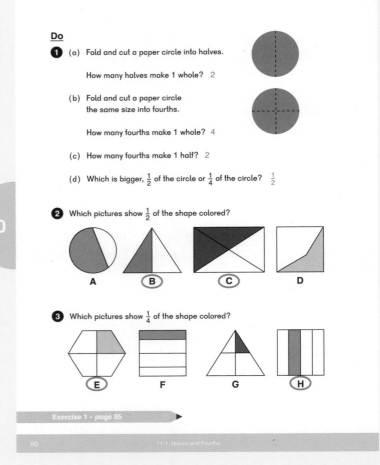

Exercise 1 • page 85

Lesson 2 Writing Unit Fractions

Objective

- Recognize and name unit fractions up to $\frac{1}{10}$.

Lesson Materials

- Sheets of square paper, several per student
- Strips of paper
- Scissors

Think

Have students discuss the questions posed in **Think**.

Provide students with square papers and pose the problems of cutting the cake into 8 equal parts.

Allow students time to experiment with ways to fold their papers to be sure that each part is equal.

Ask students how they would name and write the fraction for each piece. If they come up with the fraction $\frac{1}{8}$, you can have them label their pieces before cutting.

Learn

Have students compare their papers to the illustrations in **Learn**.

Ask students:

- How many parts do you have?
- How do we write what one part is?

Have students label each eighth as $\frac{1}{8}$ if they did not already do so. Discuss Mei's question.

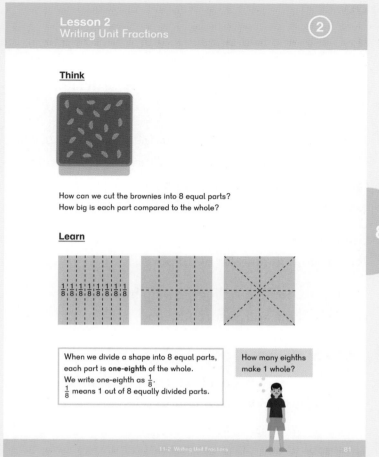

© 2017 Singapore Math Inc.   Teacher's Guide 2B Chapter 11   105

Do

① Have students fold strips of paper. To make one-sixths, fold the paper strip into thirds, then in half. For students who struggle, provide strips of paper with folding lines already marked.

If time permits, give students additional strips of paper to have them fold the strips in different ways and label for other fractions. They can save their strips for the next lesson.

Students may notice as they do this that the more parts they have, the smaller each part is.

Having students practice viewing fractions using strips will help them with bar modeling.

② Have students both say and write the fraction.

Alex's comment is an introduction to comparing fractions by looking at a unit fraction. Have students line up $\frac{1}{2}, \frac{1}{3}, \frac{1}{4}, \frac{1}{5}, \ldots \frac{1}{10}$ to see that the parts are getting smaller. Further discussion on comparing fractions will occur in Lesson 5: Comparing and Ordering Fractions.

Activity

▲ Pattern Fractions

Materials: Construction paper, geometric shapes to trace such as lids or solids from a geometry set, Dot Paper (BLM), pattern blocks, scissors

Challenge students to create other models of unit fractions using different shapes, similar to ④. They could use Dot Paper (BLM) or pattern blocks to create their shape.

Students may also create unit fractions with denominators greater than 12.

Exercise 2 • page 87

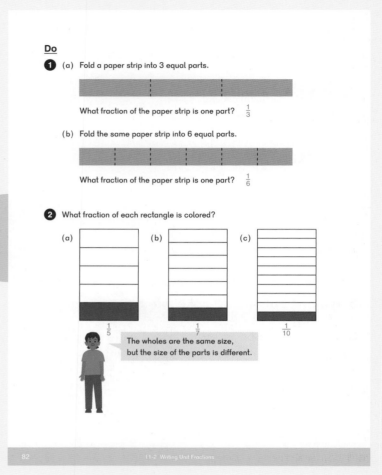

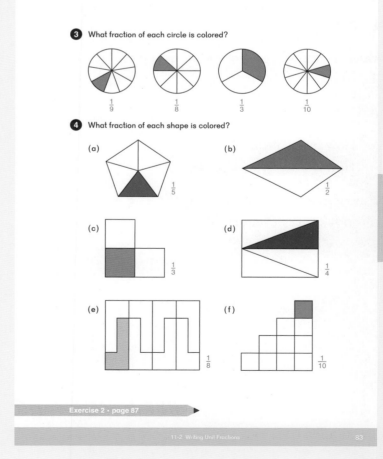

Lesson 3 Writing Fractions

Objective

- Write common fractions up to tenths.

Lesson Materials

- Strips of paper
- Fraction tiles
- Fraction circles or round paper plates
- Crayons or markers
- Dot Paper (BLM)
- Pattern blocks

Think

Provide students with paper strips and have them fold their strip into fourths and shade or color three parts.

Have students label each fourth as $\frac{1}{4}$. Discuss how $\frac{1}{4}$ and $\frac{1}{4}$ and $\frac{1}{4}$ makes three-fourths.

Learn

Have students look at textbook page 84 and compare their papers to the illustrations in **Learn**.

Emma thinks about a different way to show $\frac{3}{4}$.

Students should realize that parts of a whole in drawings do not have to be contiguous. Remind them that the top number is found by simply counting a number of parts, in this case the number that is shaded. 3 out of the 4 parts are shaded.

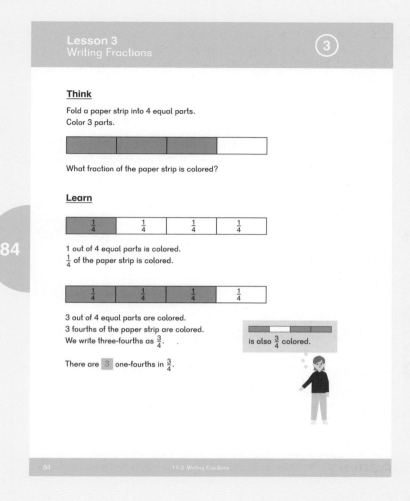

© 2017 Singapore Math Inc.　　Teacher's Guide 2B Chapter 11　　107

Do

Provide students the opportunity to show these problems with fraction tiles.

❶ Reinforce the idea that just as 1 apple and 1 apple and 1 apple and 1 apple = 4 apples, one-fifth and one-fifth and one-fifth and one-fifth is the same as four-fifths. Students may also think of this as $\frac{1}{5} + \frac{1}{5} + \frac{1}{5} + \frac{1}{5} = \frac{4}{5}$.

❷ One-eighth and one-eighth and one-eighth and one-eighth and one-eighth is the same as five-eighths. Or, $\frac{1}{8} + \frac{1}{8} + \frac{1}{8} + \frac{1}{8} + \frac{1}{8} = \frac{5}{8}$.

❹ Provide students the opportunity to show these problems with fraction circles.

❺ Challenge students to create other models of fractions using different shapes as in ❺. They could use Dot Paper (BLM) or pattern blocks to create their shape.

Activity

▲ Quilt Squares

Materials: Quilt Square (BLM), crayons or markers

Have students create a pattern on the quilt square and share what fraction of the squares and triangles they colored a certain color.

For example:

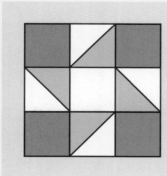

$\frac{4}{9}$ squares are red.

$\frac{4}{8}$ triangles are yellow.

Exercise 3 · page 91

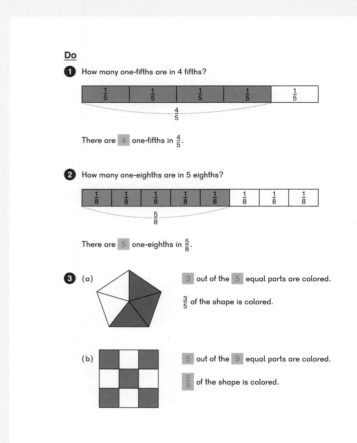

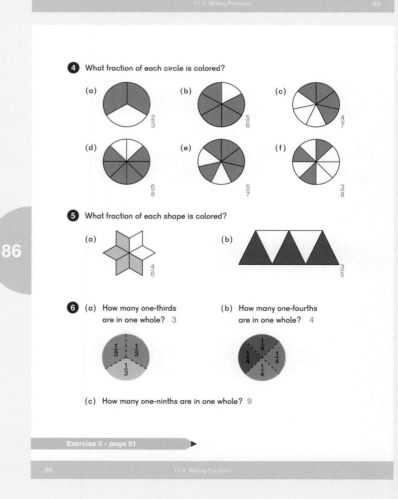

108   Teacher's Guide 2B Chapter 11   © 2017 Singapore Math Inc.

Lesson 4 Fractions that Make 1 Whole

Objective

- Find two fractions that make 1 whole.

Lesson Materials

- Fraction tiles or strips of paper

Think

Pose the **Think** problem. Have students show their answers on a whiteboard or with fraction tiles or paper strips.

Learn

Discuss the representation of the chocolate bar that was eaten and the part that is left.

Students should count up the number of pieces that were eaten and the number left to see that:

3 fifths and 2 fifths make 5 fifths, or one whole.

Or:

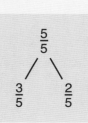

$\frac{3}{5} + \frac{2}{5} = \frac{5}{5} = 1$ whole

Provide students with other examples and have them solve with fraction tiles.

For example:

I cut another chocolate bar into 9 equal pieces and ate 3 of them. What fraction of the bar did I eat?

$\frac{1}{9} + \frac{1}{9} + \frac{1}{9} = \frac{3}{9}$

I ate $\frac{3}{9}$ of the chocolate bar.

What fraction of the bar is left?

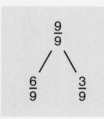

6 + 3 = 9 so...

$\frac{6}{9}$ and $\frac{3}{9} = \frac{9}{9}$, or 1 whole.

$\frac{6}{9}$ of the chocolate bar is left.

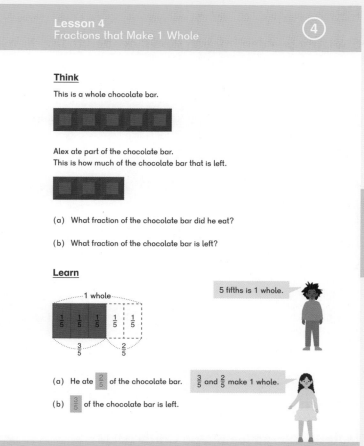

Do

③ Students can use fraction tiles if needed to find how many more parts are needed to make one whole.

Activities

● **Match**

Materials: Fraction Match Cards (BLM)

Lay cards in a faceup array. Have students match two Fraction Match Cards (BLM) with fractions that, when combined, make one whole.

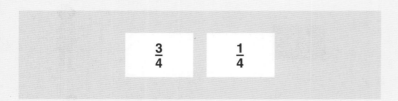

▲ **Memory**

Materials: Fraction Match Cards (BLM)

Play using the same rules as **Match**, but set the cards out facedown in an array.

▲ **The Missing Piece**

Materials: *The Missing Piece* by Shel Silverstein, circles or paper plates cut into fractional amounts and labeled with the amounts (glue googly eyes on each piece to look like pieces in the book!)

Read the book, then give each student one fractional piece of a circle. Have students find their missing piece.

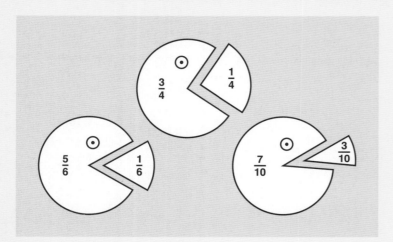

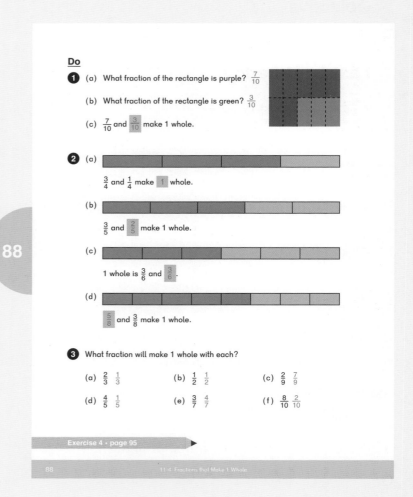

Exercise 4 • page 95

Lesson 5 Comparing and Ordering Fractions

Objective

- Compare and order fractions to tenths.

Lesson Materials

- Strips of paper, 3 per student
- Fraction tiles
- Crayons or markers

Think

Have students fold paper strips into halves, thirds, and fourths. Students can shade the first part of each strip and label each part with the correct unit fraction: $\frac{1}{2}$, $\frac{1}{3}$, and $\frac{1}{4}$.

Discuss the sizes of the unit fractions.

Ask students:

- Which fraction is the largest and which is the smallest?
- Why is $\frac{1}{4}$ smaller when the bottom number is greater than the bottom number in $\frac{1}{2}$?

Learn

Have students line their strips up to see which fraction is bigger and which is smaller.

Dion reminds students that when there are more equal parts of the same-sized whole, each part is smaller. When we have fewer equal parts of the same-sized whole, each part is larger.

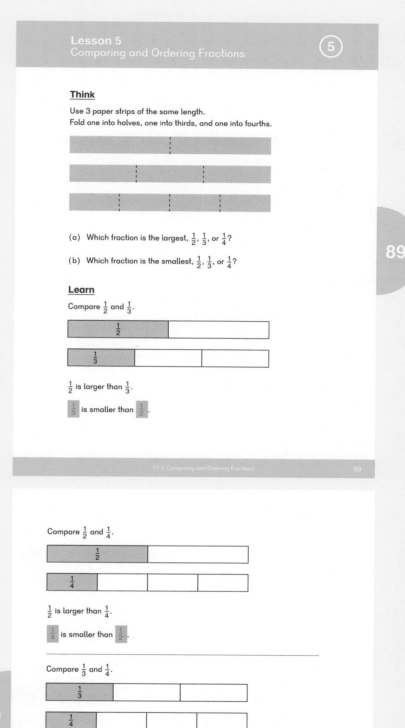

Do

❶ – ❷ Have students identify the whole first. Ask students, "If the whole is the circle, then which part is greater, $\frac{1}{3}$ of the circle or $\frac{1}{5}$ of the same circle?"

❹ Students who struggle can use fraction tiles to put the fractions in order from least to greatest.

Activities

▲ Pattern Blocks Fractions Exploration

Materials: Triangle Graph Paper (BLM), colored pencils, pattern blocks including: red trapezoid, green triangle, yellow hexagon, and blue rhombus only

Give students time to explore and discover relationships between the pattern blocks by laying them on the Triangle Graph Paper (BLM). Students could also recreate the shapes on the graph paper with colored pencils.

Ask:

- Do you see any other relationships to the (insert a shape)?
- Can you show what fractional amount the (shape) is to the (shape)?
- What shape is found in all the other shapes?

▲ Fraction Battle

Materials: Fraction Match Cards (BLM)

Playing in a group of 2 to 3 students, evenly deal out all Fraction Match Cards (BLM) facedown.

Players each flip over a card at the same time. The greatest fraction (or least depending on version of game) wins.

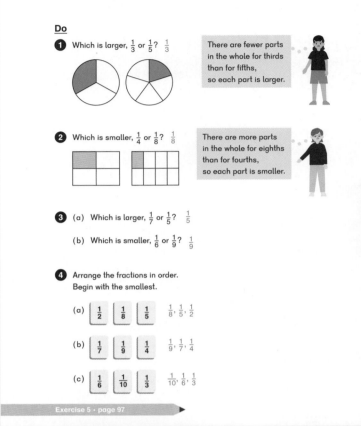

Exercise 5 • page 97

Lesson 6 Practice

Objectives

- Practice comparing and ordering fractions.
- Practice finding fractions that make 1 whole.

After students complete the **Practice** in the textbook, have them continue to practice ordering and comparing fractions with activities from the chapter.

Activity

▲ More Fraction Quilt Squares

Materials: Square pieces of blank paper, crayons or markers

Have students fold the paper to create equal parts, then color their paper quilt square with a design.

Once students have completed their quilt square, have them swap with a partner. Have their partner try to figure out and write down the fractions.

Variation: Have students divide their quilt square into eighths. Give them directions, such as:

- $\frac{4}{8}$ of the quilt square should be colored.
- $\frac{2}{8}$ of the quilt square should be striped.
- $\frac{2}{8}$ of the quilt square should be polka dotted.

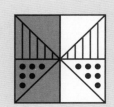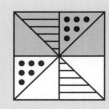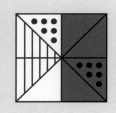

© 2017 Singapore Math Inc. Teacher's Guide 2B Chapter 11

Exercise 6 • page 101

Brain Works

★ **Pattern Blocks**

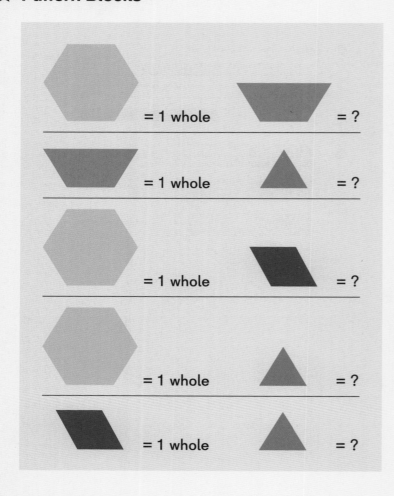

Review 3

Objective

- Cumulative review of content from Chapter 1 through Chapter 11.

Use the cumulative review as necessary to practice and reinforce content and skills from the first 11 chapters.

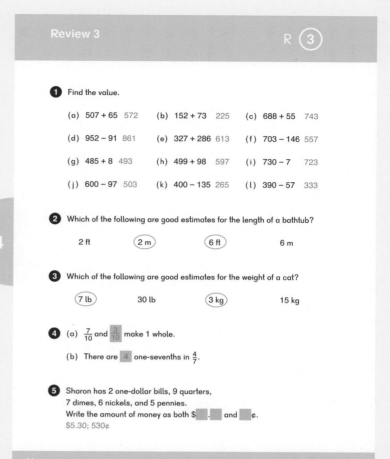

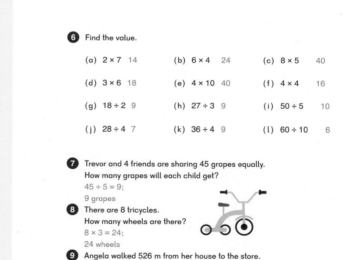

Exercise 7 • page 105

---

11 What sign, >, <, or =, goes in the ◯?

(a) 40 + 700 + 1 ⬌ > ⬌ 400 + 10 + 7
(b) 6 × 4 ⬌ > ⬌ 5 × 4 + 3
(c) 584 + 392 ⬌ = ⬌ 382 + 594
(d) 3 × 7 ⬌ < ⬌ 9 × 3
(e) 803 + 195 ⬌ > ⬌ 506 − 397
(f) 1 cm ⬌ < ⬌ 1 in
(g) 200 + 40 + 6 ⬌ = ⬌ 254 − 8
(h) $503 ⬌ > ⬌ 530¢

12 Ms. Gonzalez has 4 baskets with 10 apples in each basket.
She wants to put the apples equally into 5 bags.
How many apples should she put in each bag? 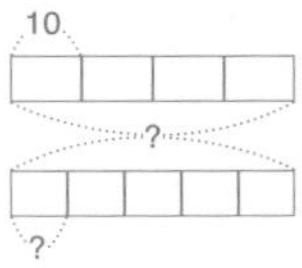
4 × 10 = 40; 40 ÷ 5 = 8;
8 apples

13 Yara had $300.
She spent $150 on a jacket and $97 on a pair of shoes.
How much money does she have left? 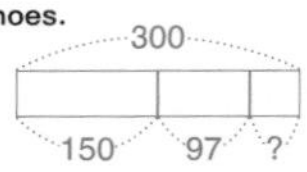
300 − 150 = 150; 150 − 97 = 53;
$53

14 Carson buys a drink that costs 75¢ and a sandwich that costs $3.35.
He pays with a 5-dollar bill.
How much change does he get?
$0.90 or 90¢

Exercise 1 • pages 85–86

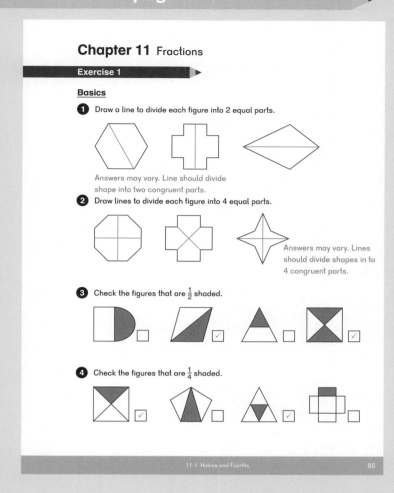
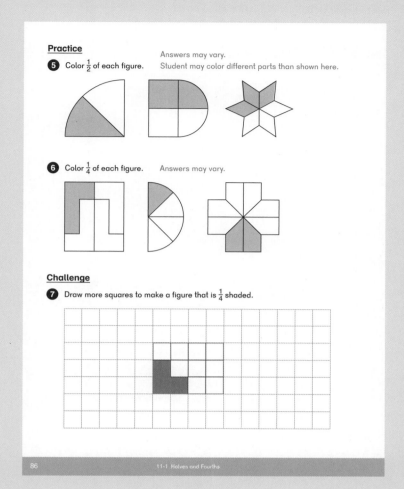

Exercise 2 • pages 87–90

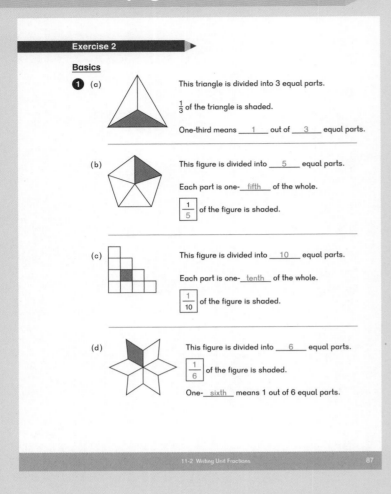

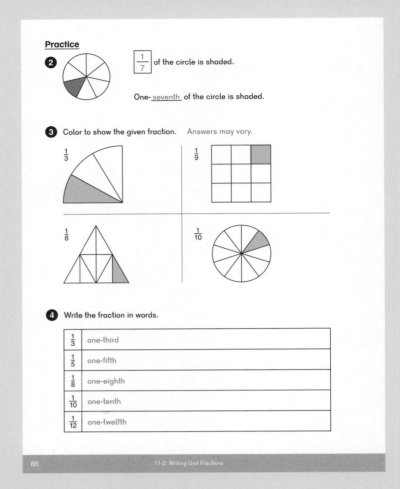

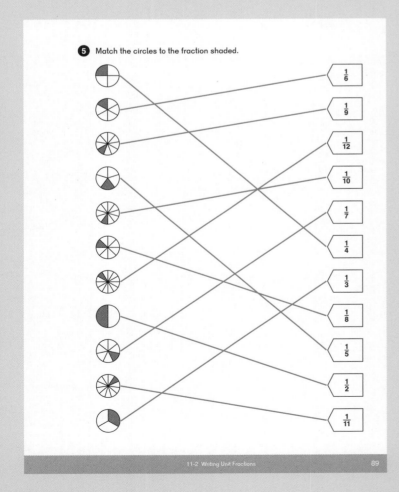

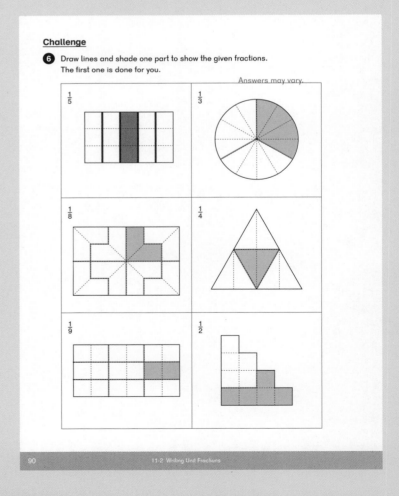

Exercise 3 • pages 91–94

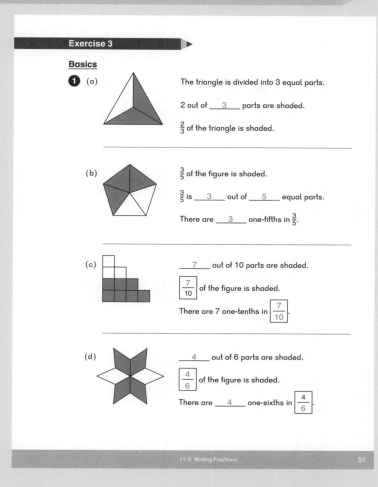

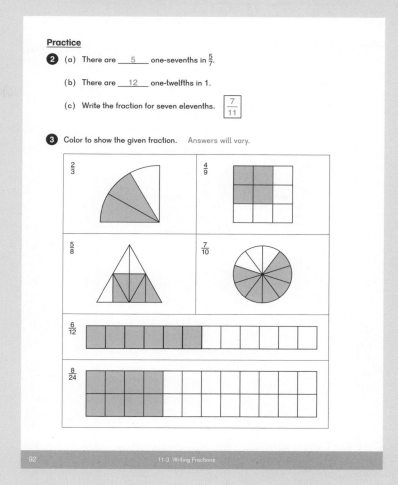

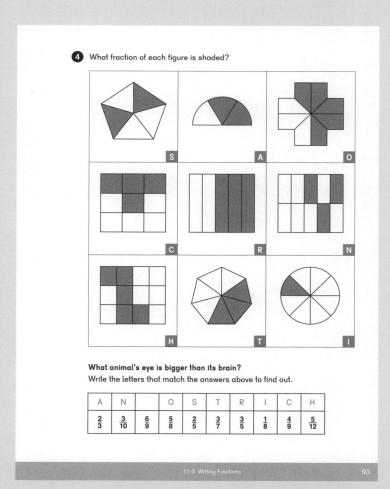

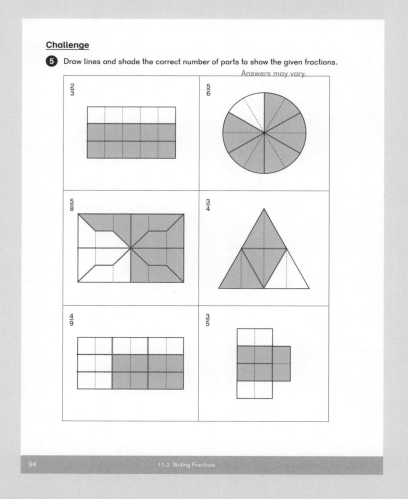

Exercise 4 • pages 95–96

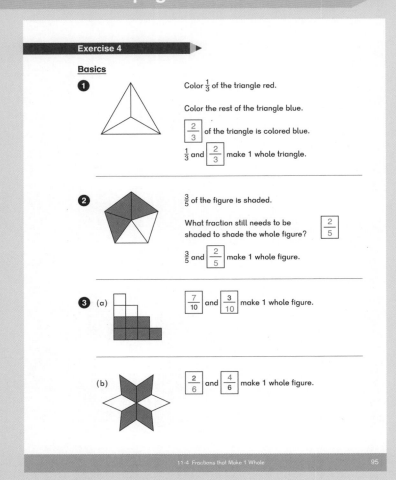
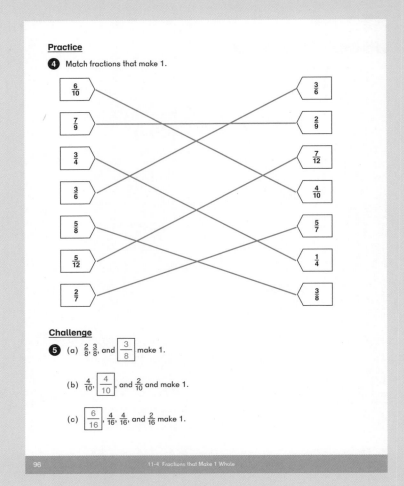

Exercise 5 • pages 97–100

Exercise 5

Basics

1. (a) $\frac{1}{3}$ of the circle is larger than $\frac{1}{4}$ of the same size circle.

   (b) $\frac{1}{5}$ of the rectangle is larger than $\frac{1}{7}$ of the same size rectangle.

2. (a) $\frac{1}{9}$ of the square is smaller than $\frac{1}{6}$ of the same size rectangle.

   (b) $\frac{1}{8}$ of the triangle is smaller than $\frac{1}{4}$ of the same-size triangle.

Practice

3. Color one part of each rectangle to show the given fraction. Then use the rectangles to answer the problems below.

   $\frac{1}{3}$, $\frac{1}{5}$, $\frac{1}{6}$, $\frac{1}{8}$, $\frac{1}{9}$, $\frac{1}{10}$, $\frac{1}{12}$

4. Circle the largest fraction.

   (a) $\frac{1}{8}$ $\left(\frac{1}{3}\right)$ $\frac{1}{10}$ (b) $\left(\frac{1}{6}\right)$ $\frac{1}{8}$ $\frac{1}{9}$

   (c) $\left(\frac{1}{5}\right)$ $\frac{1}{12}$ $\frac{1}{9}$ (d) $\frac{1}{10}$ $\frac{1}{6}$ $\left(\frac{1}{3}\right)$

5. Circle the smallest fraction.

   (a) $\frac{1}{8}$ $\frac{1}{6}$ $\left(\frac{1}{12}\right)$ (b) $\frac{1}{6}$ $\frac{1}{3}$ $\left(\frac{1}{9}\right)$

   (c) $\frac{1}{5}$ $\frac{1}{8}$ $\left(\frac{1}{9}\right)$ (d) $\left(\frac{1}{10}\right)$ $\frac{1}{6}$ $\frac{1}{5}$

6. Write the fractions in order, beginning with the smallest.

   $\frac{1}{6}$ $\frac{1}{9}$ $\frac{1}{7}$ $\frac{1}{2}$ $\frac{1}{10}$

   $\frac{1}{10}, \frac{1}{9}, \frac{1}{7}, \frac{1}{6}, \frac{1}{2}$

7. Write the fractions in order, beginning with the largest.

   $\frac{1}{4}$ $\frac{1}{12}$ $\frac{1}{5}$ $\frac{1}{8}$ $\frac{1}{11}$

   $\frac{1}{4}, \frac{1}{5}, \frac{1}{8}, \frac{1}{11}, \frac{1}{12}$

8. Darryl ate $\frac{1}{6}$ of a pizza. His brother ate $\frac{1}{3}$ of the same pizza. Who ate less?

   Darryl

9. Fang painted $\frac{1}{8}$ of a room, Debra painted $\frac{1}{5}$ of the room, and Alice painted $\frac{1}{3}$ of the room. Who painted the most?

   Alice

10. Wainani has finished reading about a third of her book. Has she finished more or less than half of her book?

    Less

Challenge

11. Use different colors to show each fraction. The colored parts should not overlap. Circle the largest fraction for each.

    Answers shown by coloring will vary.

    $\frac{1}{8}$ $\left(\frac{1}{4}\right)$ $\frac{1}{4}$ $\left(\frac{1}{3}\right)$ $\frac{1}{6}$

    $\left(\frac{1}{6}\right)$ $\frac{1}{18}$ $\frac{1}{9}$ $\frac{1}{3}$ $\left(\frac{1}{4}\right)$ $\frac{1}{6}$ $\frac{1}{12}$

12. Is two-fourths of a shape larger than two-fifths of that same shape?

    Yes. Since one-fourth is larger than one-fifth, 2 one-fourths will be larger than 2 one-fifths.

13. Circle the smaller fraction.

    (a) $\left(\frac{2}{8}\right)$ $\frac{2}{4}$ (b) $\left(\frac{2}{6}\right)$ $\frac{2}{3}$

Exercise 6 • pages 101–104

Exercise 6

Check

1. Color to show the given fraction.   Answers will vary.

   $\frac{5}{6}$, $\frac{2}{5}$, $\frac{5}{8}$, $\frac{9}{12}$, $\frac{10}{18}$, $\frac{7}{10}$

2. What fraction of each bar is shaded?

   (a) $\frac{1}{6}$
   (b) $\frac{2}{3}$
   (c) $\frac{5}{12}$
   (d) $\frac{3}{4}$
   (e) $\frac{5}{8}$
   (f) $\frac{4}{9}$

3. $\frac{3}{5}$ and $\boxed{\frac{2}{5}}$ make 1.

   $\boxed{\frac{5}{10}}$ and $\frac{5}{10}$ make 1.

   ___8___ one-eighths make 1.

   There are ___4___ one-ninths in $\frac{4}{9}$.

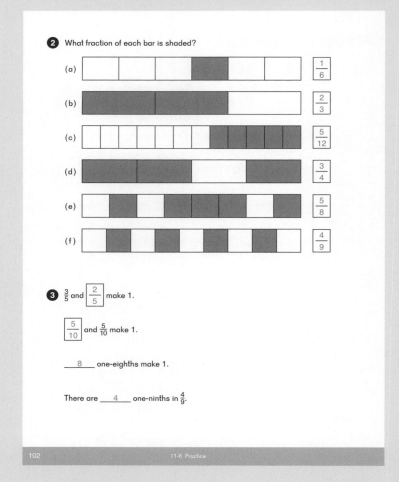

4. Circle the smallest fraction.
   (a) $\frac{1}{5}$ $\frac{1}{6}$ $\boxed{\frac{1}{7}}$
   (b) $\boxed{\frac{1}{9}}$ $\frac{1}{8}$ $\frac{1}{2}$

5. Write the fractions in order, beginning with the smallest.

   $\frac{1}{3}$ $\frac{1}{6}$ $\frac{1}{4}$ $\frac{1}{9}$

   $\frac{1}{9}, \frac{1}{6}, \frac{1}{4}, \frac{1}{3}$

6. Karen cut a pie into 8 equal pieces. Her family ate 3 pieces.

   (a) What fraction of the pie did her family eat?
   $\frac{3}{8}$

   (b) What fraction of the pie is left?
   $\frac{5}{8}$

7. Eli read about one-fifth of a book on Tuesday. He read about one-eighth of the book on Thursday. On which day did he read more of the book?

   Tuesday

Challenge

8. $\frac{1}{9}$ of the beads in a box are blue, $\frac{2}{9}$ of the beads are red, $\frac{3}{9}$ of the beads are green, and the rest are white.
   What fraction of the beads are white?

   $\frac{3}{9}$ of the beads are white.

9. Carlos used $\frac{1}{2}$ of his string and Luke used $\frac{1}{4}$ of his string to tie a package. Luke used more string than Carlos.
   Explain how this can be true.

   Luke's string was more than twice as long as Carlos' string.

10. The figure below was made by overlapping 3 squares of the same size. What fraction of the figure is shaded?

    $\frac{2}{10}$

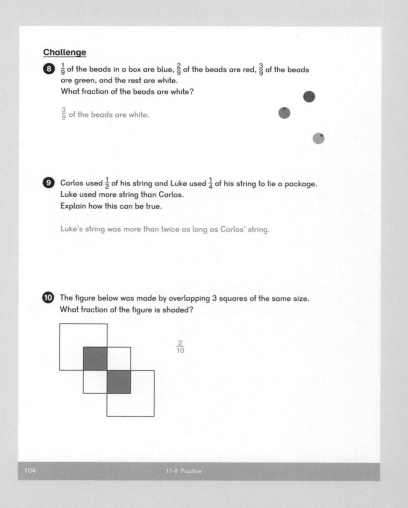

Exercise 7 • pages 105–110

Exercise 7

Check

1. (a) 7 hundreds, 6 tens, and 5 ones make __765__.
   (b) 4 hundreds, 5 ones, and 0 tens make __405__.
   (c) 20 less than 3 hundreds 1 ten and 8 ones is __298__.
   (d) __260__ is 5 tens less than 3 hundreds 1 ten.

2. (a) 354 + 80 = __434__
   (b) 432 − 97 = __335__
   (c) 64 + __36__ = 100
   (d) 600 + 4 + 50 = __724__ − 70
   (e) 4 × 3 = __24__ ÷ 2

3. Write >, <, or = in the ◯.
   (a) 500 + 120 + 63 __<__ 80 + 503 + 200
   (b) 358 + 98 __>__ 453 − 97
   (c) 60 + 70 + 80 + 90 __>__ 120 + 50 + 40 + 80
   (d) 5 × 2 __<__ 3 × 4

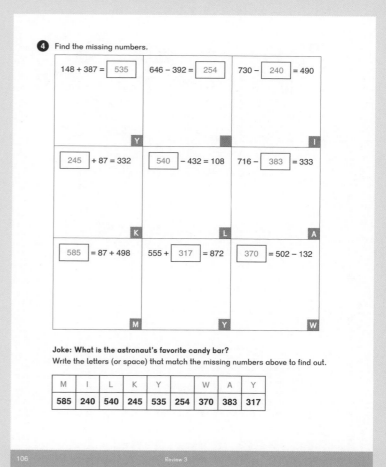

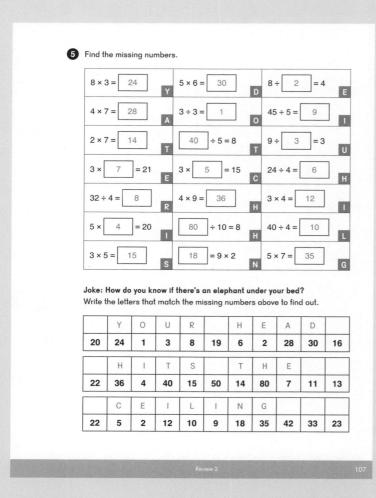

⓫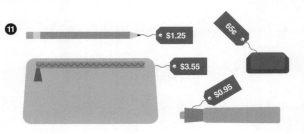

(a) Lincoln has two quarters and a dime.
How much more money does he need to buy the pencil?

$1.25 − $0.60 = $0.65
He needs $0.65 more.

(b) How much more does the pencil case cost than the eraser?

$3.55 − $0.65 = $2.90
It costs $2.90 more.

(c) Rowan bought the pencil case.
She paid with a five-dollar bill.
How much change did she get?

$5.00 − $3.55 = $1.45
She got $1.45 change.

(d) Jamal bought all 4 items.
How much did he spend?

$1.25 + $3.55 + $0.65 + $0.95 = $6.40
He spent $6.40.

Challenge

⓬ The shaded part of each square is marked with a value.
What is the value of the whole square?

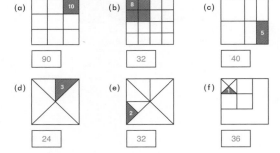

(a) 90 (b) 32 (c) 40

(d) 24 (e) 32 (f) 36

⓭ There are 4 ropes, A, B, C, and D.
C is longer than A but shorter than B.
D is longer than B.
Match the ropes with their lengths.

Students should know from measuring using these units that
1 cm < 1 in < 1 ft < 1 m,
from which they can extrapolate that
20 cm < 20 in < 20 ft < 20 m.

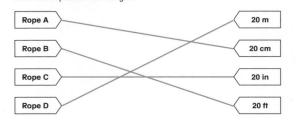

Rope A — 20 m
Rope B — 20 cm
Rope C — 20 in
Rope D — 20 ft

Chapter 12 Time   Overview

Suggested number of class periods: 4–5

	Lesson	Page	Resources	Objectives
	Chapter Opener	p. 127	TB: p. 97	Investigate telling time.
1	Telling Time	p. 128	TB: p. 98 WB: p. 111	Tell time to the nearest minute. Relate a time on the clock to the amount of time that has passed between a time on the hour ("o'clock") and a time to the minute within that hour.
2	Time Intervals	p. 131	TB: p. 103 WB: p. 115	Determine elapsed time in minutes within 1 hour, and in hours, given the start time and end time. Tell the end time given the start time and the elapsed time.
3	A.M. and P.M.	p. 134	TB: p. 107 WB: p. 121	Tell time using a.m. and p.m.
4	Practice	p. 136	TB: p. 110 WB: p. 125	Practice telling time.
	Workbook Solutions	p. 138		

© 2017 Singapore Math Inc.   Teacher's Guide 2B Chapter 12

Chapter 12 Time — Notes & Materials

In **Dimensions Math® 1B** Chapter 18: Time, students learned:

- To identify the hour and minute hands.
- To tell time on digital and analog clocks to the half hour, quarter hour, and 5-minute marks.
- That there are 60 minutes in 1 hour.

In this chapter, students will build on their knowledge from **Dimensions Math® 1B** by learning to tell time to the minute. In addition, students will learn to find elapsed time in minutes within 1 hour, and in hours given the start time and end time.

Dimensions Math® 3 will focus on intervals of elapsed time across a.m. and p.m.

It is important for students to see that the clock hands do not move independently, but together. Therefore, use a demonstration clock with geared hands where the hour hand moves in conjunction with the minute hand, not separately. (Often called "Judy Clocks.")

Teachers should incorporate time into daily activities.

For example:

- It's time to start math class, what time is it?
- We are going to gym, what time is it?
- Math class ends in one hour. What time will it be when math class ends?
- How much time before we go to recess?

Materials

- Chalk
- Dry erase markers
- Dry erase sleeves
- Glue
- Painter's tape
- Paper plates
- Recording sheet listing all students in the class
- Strips of paper
- Student clocks with geared hands
- Demonstration clock
- Whiteboards

Blackline Masters

- 5-Minute Time Match Cards
- Clock Face
- Time Cards
- Watch Face

Storybooks

- *Around the Clock* by Roz Chast
- *A Second, a Minute, a Week with Days in It* by Brian P. Cleary
- *I.Q. It's Time* by Mary Ann Frazer
- *Telling Time with Big Mama Cat* by Dan Harper
- *The Clock Struck One* by Trudy Harris
- *A Second is a Hiccup* by Hazel Hutchings
- *What Time Is It, Mr. Crocodile?* by Judy Sierra
- *Me Counting Time: From Seconds to Centuries* by Joan Sweeney

Chapter Opener

Objective
- Investigate telling time.

Lesson Materials
- Student clocks with geared hands
- Demonstration clock

Discuss the different times that the friends are doing their activities on Sports Day.

Guide students in counting by five minute increments from the 12 to determine times to the 5-minute mark. They can also count on from a known time, such as counting on from half past, or 30 minutes past: 35, 40, ... Demonstrate how the hands move and write the times on the board.

Remind students that the time for the hour is followed by a colon, then the time for the minutes, which is always two digits. For example, 9:05 is 5 minutes after 9.

Have students show times on their clocks. Discuss the hour and minute hands and how they move together. Ask students what time they do things at home such as eat dinner, wake up, etc.

Students learned about fractions in the previous chapter. Have them see, by moving the minute hand of the clock, that as the minute hand moves from 0 to 30, it moves one-half of the way around the clock, and as it moves from 0 to 15, it moves one-fourth, or one quarter, of the way around the clock.

Although students have not yet learned to find fractions of quantities, they may realize that 15 minutes is one-fourth of 60 minutes, or one-fourth of an hour.

Chapter 12

Time

Our friends are having fun on Sports Day.
What time are they doing each activity?

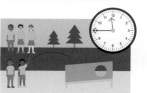

Activities

● Match

Materials: 5-Minute Time Match Cards (BLM)

Lay cards in a faceup array. Each student chooses a card and looks for the matching time.

▲ Memory

Materials: 5-Minute Time Match Cards (BLM)

Play using the same rules as **Match**, but set the cards out facedown in an array.

Lesson 1 Telling Time

Objectives

- Tell time to the nearest minute.
- Relate a time on the clock to the amount of time that has passed between a time on the hour ("o'clock") and a time to the minute within that hour.

Lesson Materials

- Student clocks with geared hands
- Demonstration clock

Think

Pose the **Think** problem about Mei's math class. Provide students with clocks and have them work through the problem. Ask them how they determined the time.

From previous experience with telling time to the 5-minute mark, some students may have counted by fives to 50, then by ones to 53. Others might have counted on from the half hour or the three-quarter hour mark. Some students might even have counted backwards from 60.

Learn

Have students set their clocks to 10:00 and move them through the first four examples in **Learn**: 10:05, 10:15, 10:30, and 10:45.

At each time, discuss:

- How the time is said aloud: "ten oh five," "half-past ten," "10:30," etc.
- How much time has passed since 10:00.
- How they can skip count by fives to find the correct time.

Students should also note how the hour hand moves from one number (10) to the next (11).

Note that students can be confused by which hour to name when the hour hand has advanced to near the

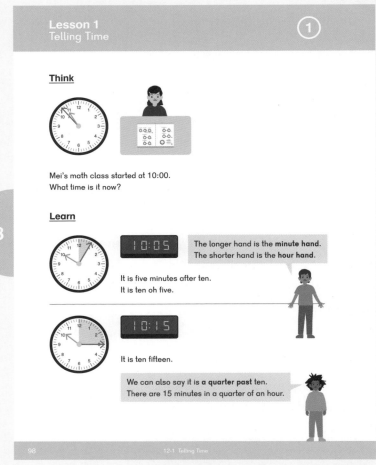

next hour. Be sure to emphasize it's *almost* to the next hour, therefore, it's not yet that hour.

For example, at 10:55, the hour hand is very close to 11:00, but until the minute hand passes the hour, the time is still read as the previous hour.

128   Teacher's Guide 2B Chapter 12   © 2017 Singapore Math Inc.

At 10:53, Emma points out that we can skip count by fives and then add 3 to get to 53 minutes past 10:00.

Provide additional examples of time to the minute on a demonstration clock. Ask students to tell the time until they are comfortable reading time to the 1-minute mark.

Include a few examples where you ask students how many minutes it is before the hour.

Do

Students can use small student clocks for all of the questions on pages 100 and 101 so they can physically move the minute hand as they count by fives and then by ones, and see the movement of the hour hand. Some students may simply count using the clock pictures, but using a clock will give them a better sense of how the time changed.

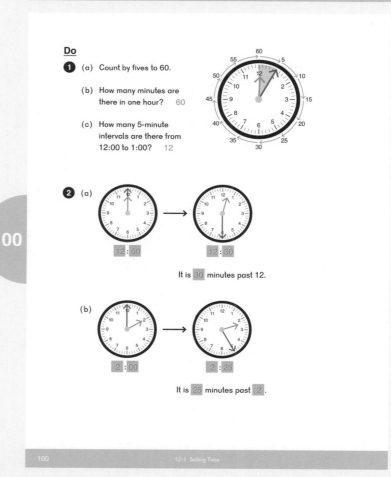

3) Have students tell you the time after the hour for each of these problems. Provide further practice by having students go back to 2) on page 100 and tell you the times to the minute.

3) — 4) These problems are designed to give students a sense of time passing, from waking up and getting to school. Take the opportunity to point out the time on the classroom clock regularly to help students get a sense of passing time.

Activity

▲ Watch

Materials: Watch Face (BLM), Time Cards (BLM), strips of paper, glue, recording sheet with names of all students in the class for each student

Have each student draw (pulling from a hat is fun) a Time Card (BLM) showing a digital time.

Have students draw hands on the Watch Face (BLM) to match their digital time.

Students can then staple or tape the watch together to wear on their wrists.

Have students ask classmates, "What time is it?" The classmate being asked will show his watch to the student asking for the time, who will then write down the time shown on his classmate's watch next to that classmate's name.

Exercise 1 • page 111

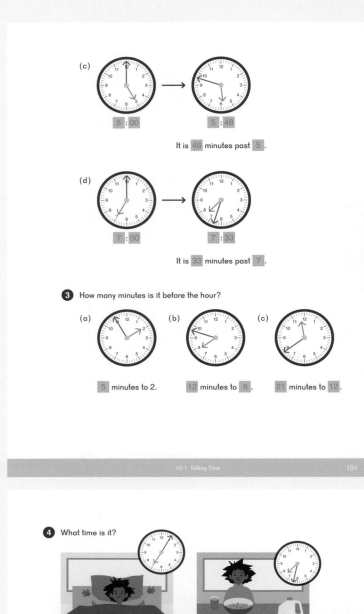

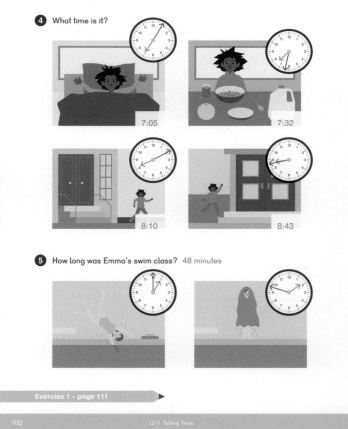

Lesson 2 Time Intervals

Objectives

- Determine elapsed time in minutes within 1 hour, and in hours, given the start time and end time.
- Tell the end time given the start time and the elapsed time.

Lesson Materials

- Student clocks with geared hands
- Demonstration clock
- Clock Face (BLM)
- Dry erase markers
- Dry erase sleeves

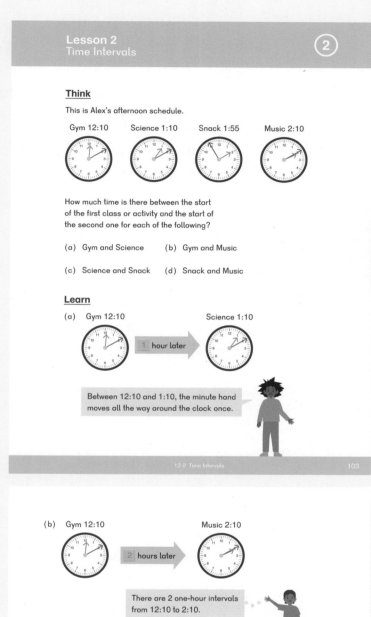

Think

Pose the questions of time between activities. Provide students with geared clocks to work the problems. Have students share and discuss their strategies for finding the time that has passed. In particular, they should also note when the minute hand again passes the 12 for the start of a new hour.

Learn

Students should set their clocks to the first time and move the minute hand to the next time, counting the minutes or hours.

From gym to science they should notice the hour hand moved 1 hour, so no counting is needed. From gym to music it's 2 hours. Ask students, "From gym to science, what changed on the clock?"

For (a) and (b), ask students, "Which hand ends up in a different place at the end of the time interval?"

Question (d) is the first example where the elapsed time is in minutes and the time changes from one hour to the next. Have students discuss how they found the time.

After working the problems, provide additional practice with the clocks.

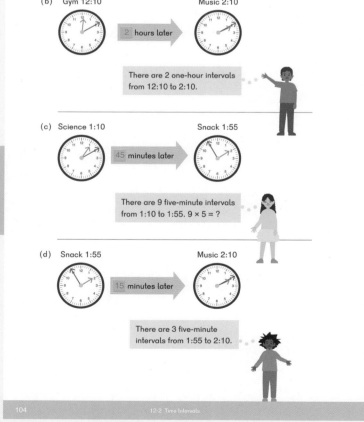

© 2017 Singapore Math Inc.   Teacher's Guide 2B Chapter 12   131

Do

Students should continue to use clocks as needed.

3 – 4 Students count on or back in 5-minute increments.

Students can show their answers with the Clock Face (BLM) in a dry erase sleeve.

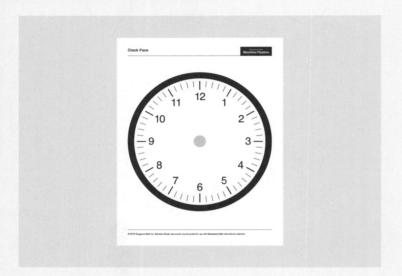

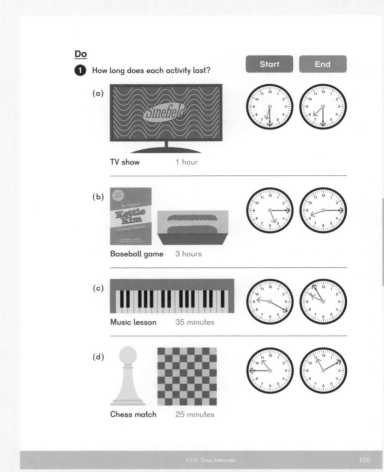

Activity

▲ **Elapsed Time Hop**

Materials: Chalk or paper plates and painter's tape

Using chalk or paper plates and tape, create a large clock face.

Have one student act as the hour hand and another as the minute hand. Give students a start time to show and then a time interval. Students find the end or start time.

For example, you say, "Show me 6:10." The hour hand student stands on 6 and the minute hand student stands on 10.

Then ask, "What time is it 2 hours later?" The student on the hour hand walks clockwise, following the numbers on the clock, to the new time and says the time (8:10).

Or, from 6:10, ask, "What time is it 15 minutes earlier?" The student representing the minute hand will walk back (counterclockwise) 15 minutes. As she crosses 12:00, the student representing the hour hand should move as well. Students should say, "The time is 5:55."

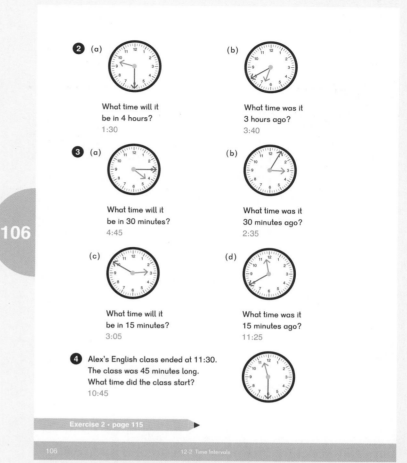

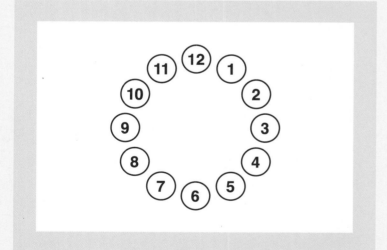

Exercise 2 • page 115

Lesson 3 A.M. and P.M.

Objective

- Tell time using a.m. and p.m.

Think

Have students look at **Think** and answer the questions. The two times are the same, so they should notice that the time of day is different from one illustration to the next to answer the second question.

See if students can come up with other examples where the time is the same but the time of day is different. For example, they might get to school at the same clock time they go to bed.

Learn

Students should notice that there are 24 hours in one day, but only 12 hours on the clock. Explain that we use the terms **a.m.** and **p.m.** to tell whether a time is between midnight and noon or noon and midnight. Clocks only have 12 hours.

Discuss the textbook example with the timeline.

Most digital clocks designate midnight as 12 a.m. and noon as 12 p.m. The terms **midnight** and **noon** are generally considered less confusing designations and will be used in this series.

Note: Students may bring up the idea of a 24-hour clock, also referred to as international time or military time. In 24-hour notation, the day begins at midnight or 00:00 and time is read in the hundreds. A 24-hour clock is commonly used in the sciences, medical fields, and airlines to ensure there is no confusion between a.m. and p.m. times. The colon or period is used in some countries and fields, and omitted in others.

8:00 a.m. can be written 08:00, 0800, or 08.00 and is read, "Oh eight hundred."

8:00 p.m. can be written 20:00, 2000, or 20.00 and is read, "Twenty hundred."

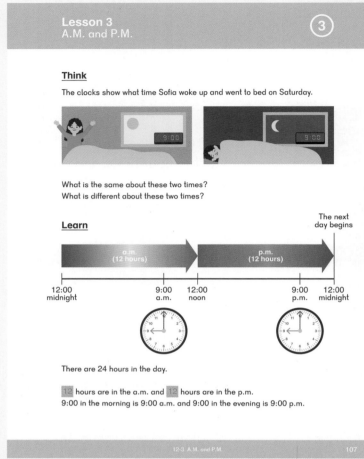

Do

③ Students should write the start time of 11:55 a.m. and the end time of 12:45 p.m. for Dion's music class.

Students should note that the time moves from a.m. to p.m. They may solve by first thinking that 11:55 to 12:00 is 5 minutes, then 12:00 to 12:45 is 45 minutes, and adding the minutes together. 5 + 45 = 50 minutes.

Students should realize that on this kind of clock (analog clock) there is no way to tell from the clock alone whether the time is a.m. or p.m.

④ Note that on a digital clock, the a.m. and p.m. are included.

Activity

▲ Story Time

Materials: Clock Face (BLM), dry erase markers, dry erase sleeves, storybooks that mention time (see suggestions on page 126 of this Teacher's Guide)

Whenever reading or telling a story that mentions time, have students show the hands on the Clock Face (BLM) in dry erase sleeves. For example, "It was Wednesday morning and Alex had to go to school."

- What time do you think he woke up?
- What time do you think he had to be at school?
- How long did he have between the two times?

Students can share their ideas and make the time on their clocks.

This can also be done with any project or event that's happening.

Exercise 3 • page 121

Do

① Write what time you usually do these activities. Answers may vary.
Include a.m. or p.m.

(a) Wake up.
(b) Eat breakfast.
(c) Leave for school.
(d) Eat lunch.
(e) Leave school.
(f) Eat dinner.
(g) Do your homework.
(h) Go to bed.

② Write what time Mei does each activity.
Include a.m., p.m., noon, or midnight.

(a) 7:12 a.m.
(b) 8:04 a.m.
(c) 12:00 noon
(d) 4:15 p.m.
(e) 8:56 p.m.
(f) 12:00 midnight

③ The clocks show when Dion's music class begins and ends.

Begin   End

(a) Write the start and end times. Include a.m. or p.m.
Start: 11:55 a.m., End: 12:45 p.m.

(b) How long is his music class?
50 minutes

④ The clocks show when Sofia's baseball game begins and ends.

Begin   End

How long was her game? 3 hours

⑤ A tour began at 11:00 a.m. and lasted 5 hours.
What time did it end? 4 p.m.

Exercise 3 • page 121

Lesson 4 Practice

Objective

- Practice telling time.

After students complete the **Practice** in the textbook, continue to look for situations throughout the school year where students can practice determining elapsed time during the school day.

Examples:

- We have an assembly at 10:10 and it should last one hour. When will the assembly end?
- The buses arrive at 2:50 and leave school at 3:05. How long are the buses at school loading students?

Activity

▲ Clock Nim

Materials: Student clocks with geared hands

Beginning with the hands on the clock at 12:00, players take turns moving the minute hand clockwise 5, 10, or 15 minutes.

The player who moves the hands to 3:00 exactly is the winner.

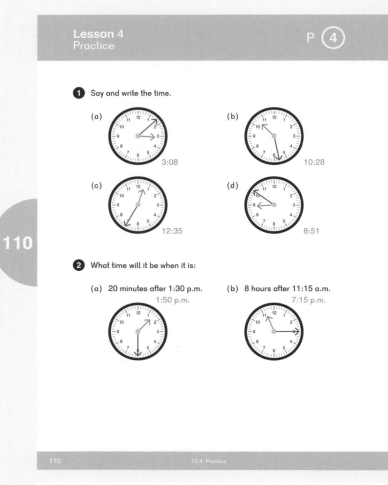

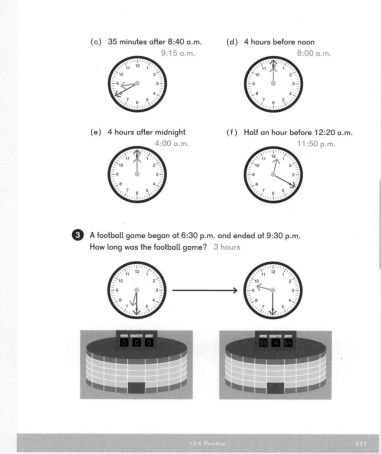

Exercise 4 • page 125

Brain Works

★ **How Long?**

A team rode a bike relay race. Each person rode one leg of the race.

- Olga started her leg at 1:00 p.m. and rode for 2 hours and 30 minutes.
- Emily rode from 9:15 a.m. until 11:30 a.m.
- Carlos rode the leg between Emily and Olga.
- Jamal rode after Olga until 4:45 p.m.

(a) Who rode the greatest amount of time?

(b) Who rode for the least amount of time?

(c) How long was the team riding in all?

Emily: 9:15 a.m. — 11:30 a.m., 2 hours and 15 minutes
Carlos: 11:30 a.m. — 1 p.m., 1 hour and 30 minutes
Olga: 1 p.m. — 3:30 p.m., 2 hours and 30 minutes
Jamal: 3:30 p.m. — 4:45 p.m., 1 hours and 15 minutes

(a) Emily

(b) Jamal

(c) 7 hours and 30 minutes

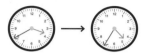

④ Pedro's band practice begins at 3:40 p.m. and ends at 4:35 p.m. How long is his band practice? 55 minutes

⑤ Sarah's computer class begins at 10:30 a.m. and lasts 40 minutes. What time does her computer class end?
11:10 a.m.

⑥ A concert began at 5:35 p.m. and lasted 4 hours. What time did the concert end?
9:35 p.m.

⑦ A 45-minute show ended at 12:15 p.m. What time did the show start?
11:30 a.m.

⑧ A 2-hour soccer match ended at 3:20 p.m. What time did the match begin?
1:20 p.m.

Exercise 4 • page 125

Exercise 1 • pages 111–114

Chapter 12 Time

Exercise 1

Basics

① Write the times and fill in the blanks.

 It is __2__ o'clock.

2 : 00

It is __40__ minutes past 2.
It is __20__ minutes to 3.
It is two forty.

2 : 40

It took 40 minutes for the minute hand to move from the 12 to the __8__.

It is __43__ minutes past 2.
Write the time in words: __two forty-three__.

2 : 43

The hour hand is between 2 and __3__.
The minute hand is between __8__ and __9__.
In another __17__ minutes it will be 3 o'clock.

Practice

② Cross out the clock that cannot be correct.

③ Which clock correctly shows 4:32?

☐   ✓   ☐

④ These clocks show different times in the morning. Write the times in order from earliest time to latest time.

6:15, 6:35, 6:49, 7:10

⑤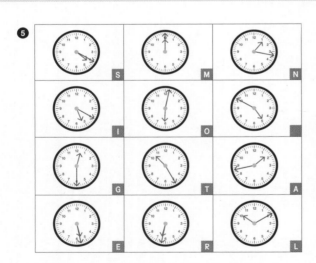

This poisonous desert-dwelling lizard from the American Southwest can live for up to 30 years.
Write the letters (or space) that match the times above to learn which animal it is.

G	I	L	A		M	O	N	S	T	E	R
12:30	5:20	10:10	1:43	4:50	12:00	6:02	1:17	4:20	10:25	5:28	6:32

⑥ Draw the minute hand on each clock face to show the time.

(a)
8:35

(b)
3:17

(c)
Half past 10

(d) A quarter to 4

(e)
seven oh seven

(f)
8 minutes to 12

Challenge

⑦ Draw the minute hand and hour hand on each clock face to show the time.

(a)
9:45

(b)
2:30

Position of hour hand can be approxiamte, but should be a bit before 10 for (a) and halfway between 2 and 3 for (b).

Exercise 2 • pages 115–120

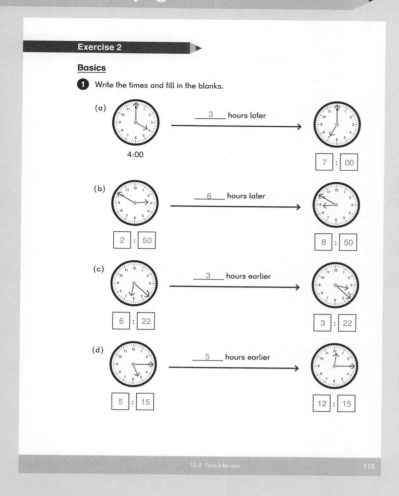

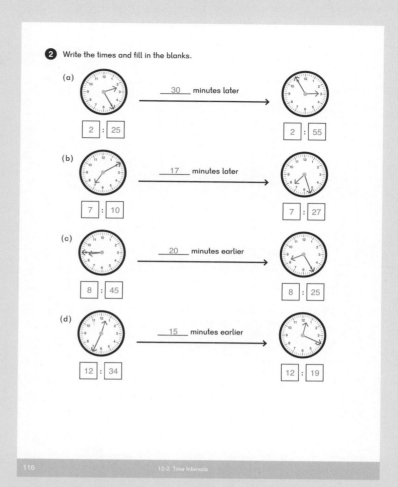

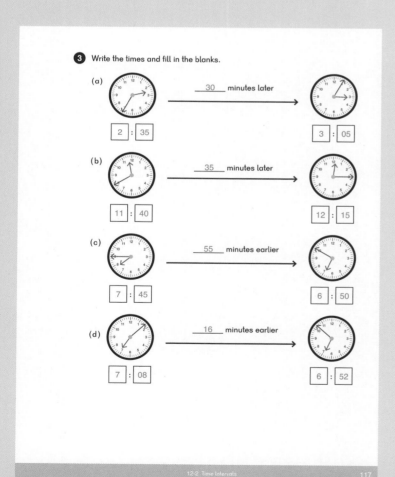

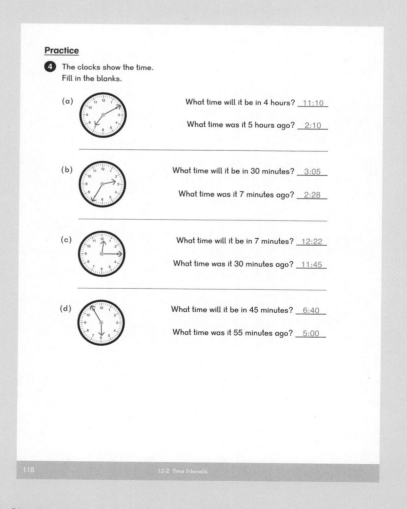

Teacher's Guide 2B Chapter 12

5. Aliyah's piano lesson started at 3:30.
   The lesson was 45 minutes long.
   What time did it end?

   4:15

6. A concert lasted 2 hours.
   It ended at 11:40.
   What time did it start?

   9:40

7. A ferry goes from the mainland to an island.
   The ferry arrived at the island at 6:10.
   The crossing time was 42 minutes.
   What time did the ferry leave the mainland?

   5:28

Challenge

8. A ferry leaves at 10:55.
   Manuel wants to be in line for the ferry an hour ahead of time.
   It takes him 2 hours to drive from home to the ferry.
   What time should he leave home?

   7:55

9. Draw hands on the clock with the missing hands to complete the patterns.

   (a)

   (b)

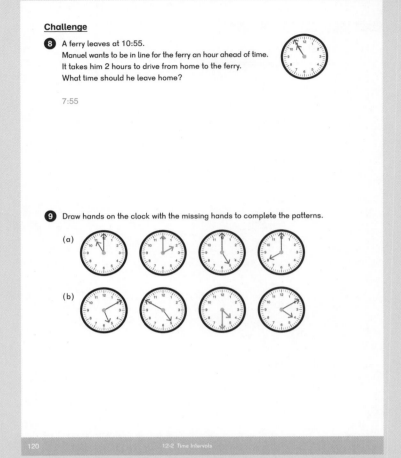

Exercise 3 • pages 121–124

Exercise 3

Basics

1 Fill in the blanks.

(a) 1 day = ⟦24⟧ hours

(b) There are __12__ hours between midnight and noon.

(c) There are __12__ hours between noon and midnight.

2 Write a.m. or p.m. in the blanks.

(a)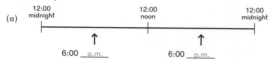

6:00 __a.m.__   6:00 __p.m.__

(b) 10:00 __p.m.__ is between noon and midnight.

(c) Jamal woke up after a night's sleep at 7:00 __a.m.__.

(d) Imani finished dinner at 6:15 __p.m.__.

(e) Evan ate lunch at 12:30 __p.m.__.

(f) A soccer game began at 11:30 __a.m.__ and ended at 12:45 __p.m.__.

(g) Kai wanted to finish all his gardening before noon, so he began at 9:00 __a.m.__.

(h) To see the night-time glowing algae in the water, Jo was on the beach at 10:00 __p.m.__.

3 Write the times using a.m. or p.m.

(a)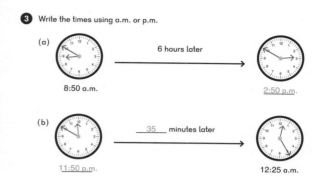

8:50 a.m.   6 hours later   __2:50 p.m.__

(b) 11:50 p.m.   __35__ minutes later   12:25 a.m.

Practice

4 The clocks show the time.
Fill in the blanks, including a.m. or p.m.

(a)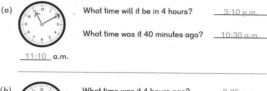

What time will it be in 4 hours?   __3:10 p.m.__
What time was it 40 minutes ago?   __10:30 a.m.__

__11:10__ a.m.

(b)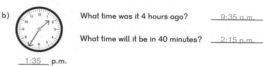

What time was it 4 hours ago?   __9:35 a.m.__
What time will it be in 40 minutes?   __2:15 p.m.__

__1:35__ p.m.

5 Write the times, including a.m. or p.m.

(a) 5 minutes after 10 in the morning   ⟦10⟧ : ⟦05⟧ __a.m.__

(b) Half past 1 in the afternoon   ⟦1⟧ : ⟦30⟧ __p.m.__

(c) A quarter to 7 in the evening   ⟦6⟧ : ⟦45⟧ __p.m.__

(d) 5 minutes to 10 in the morning   ⟦9⟧ : ⟦55⟧ __a.m.__

(e) 23 minutes after midnight   ⟦12⟧ : ⟦23⟧ __a.m.__

6 Raj's tennis lesson ended at 12:35 p.m.
The lesson was 50 minutes long.
What time did it start?

11:45 a.m.

7 Cora went on a 3-hour hike.
The hike ended at 2:40 p.m.
What time did it start?

11:40 a.m.

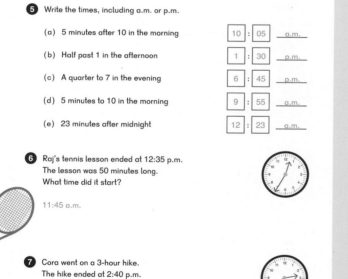

Challenge

8 There are 24 different time zones around the world.
The table below shows the times in some cities in the world when it is midnight in Honolulu.

Honolulu	12 midnight
Denver	4:00 a.m.
New York City	6:00 a.m.
London	11:00 a.m.
Moscow	1:00 p.m.
Singapore	6:00 p.m.
Auckland	10:00 p.m.

Auckland is 10 hours ahead of Honolulu, so when it is 12 midnight in Honolulu, it is 10 a.m. in Auckland.

(a) Singapore is __7__ hours ahead of London.

(b) When it is 10:00 p.m. in London, it is __5 a.m.__ in Singapore.

(c) Honolulu is __6__ hours behind New York City.

(d) When it is 10:00 p.m. in Honolulu, it is __4 a.m.__ in New York City.

(e) Moscow is __9__ hours behind Auckland.

(f) When it is 4:10 a.m. in Moscow, it is __1:10 p.m.__ in Auckland.

(g) When it is 4:10 a.m. in Auckland, it is __7:10 p.m.__ in Moscow.

Exercise 4 • pages 125–128

Exercise 4

Check

1. What time is it?

   (a)   (b)   (c)

   8:50   2:29   9:07

2. Draw the minute hands to show the times.

   (a)   (b)   (c)

   20 minutes to 4   a quarter past 1   6:18

3. (a) When the minute hand moves halfway around the clock, __30__ minutes have passed.

   (b) When the hour hand moves halfway around the clock, __6__ hours have passed.

   (c) When the minute hand moves a quarter of the way around the clock, __15__ minutes have passed.

   (d) When the hour hand moves a quarter of the way around the clock, __3__ hours have passed.

4. This clock is 6 minutes fast. What is the actual time?

   4:39

5. This clock is 12 minutes slow. What is the actual time?

   10:37

6. This clock stopped working 4 hours ago. What time is it now?

   10:05

7. Aurora wants to read for a half hour before lights off at 8:30 p.m.
   It takes her 20 minutes to get ready for bed.
   What time should she start getting ready for bed in order to be able to read for a half hour?

   7:40 p.m.

8. Write the times, including a.m. or p.m.

   (a) 4 minutes to 10 in the morning — 9 : 56 a.m.
   (b) 10 minutes past 4 in the afternoon — 4 : 10 p.m.
   (c) 2 hours after noon — 2 : 00 p.m.
   (d) Sunrise at six oh five — 6 : 05 a.m.
   (e) 4 minutes to midnight — 11 : 56 p.m.

9. Fadiya went to bed at 8:15 p.m. and slept for 9 hours. What time did she wake up?

   5:15 a.m.

10. Martin took 20 minutes to eat lunch.
    He finished lunch at 12:10 p.m.
    What time did he start lunch?

    11:50 a.m.

Challenge

11. It is 12 hours later in Singapore than it is in New York City.
    It is 6 hours earlier in Berlin than it is in Singapore.
    If it is 10:00 a.m. in New York City, what time is it in Berlin?

    4:00 p.m.

12. It is 3 hours earlier in Honolulu than in Seattle.
    An airplane flight from Seattle to Honolulu takes 6 hours.

    (a) If it is 1:00 p.m. in Seattle, what time is it in Honolulu?

    10:00 a.m.

    (b) A flight leaves Seattle at 1:00 p.m.
    What time will it be in Honolulu when the flight lands there?

    4:00 p.m.

    (c) Another flight from Seattle arrives in Honolulu at 1:00 p.m.
    What time was it in Seattle when the flight took off from there?

    10:00 a.m.

13. Some countries do not use a.m. and p.m.
    They use a 24-hour clock, instead of a 12-hour clock.
    1:00 p.m. in 12-hour time is 13:00 in 24-hour time.
    What time is it in 24-hour time at 6:00 p.m.?

    18:00

Chapter 13 Capacity

Overview

Suggested number of class periods: 3–4

	Lesson	Page	Resources	Objectives
	Chapter Opener	p. 145	TB: p. 113	Investigate capacity.
1	Comparing Capacity	p. 146	TB: p. 114 WB: p. 129	Understand the meaning of capacity. Compare the volume of water in two or more containers by direct and indirect comparison.
2	Units of Capacity	p. 149	TB: p. 118 WB: p. 133	Measure capacity in liters, quarts, and gallons.
3	Practice	p. 152	TB: p. 123 WB: p. 137	Practice measuring and comparing capacity.
	Workbook Solutions	p. 153		

Chapter 13 Capacity

Notes & Materials

In **Dimensions Math® 2A** students learned to measure and compare units of length and weight:

- Centimeters and meters
- Inches and feet
- Grams and kilograms
- Pounds

In this chapter, students extend their knowledge of measuring to understanding capacity in liters, gallons, and quarts.

It is important for students to experience capacity through hands-on activities using containers and water to help them develop quantitative/measurement sense.

The capacity of a container is only an approximation, as generally the liquid does not fill it completely, but rather to a fill line. However, that is sufficient to compare capacities (or size) of different containers.

Much of the work involves filling containers with water. 1-liter bottles with the tops cut off and labels removed make good common-sized containers.

Students will also learn to estimate the capacity of containers using liters, gallons, and quarts. Although both metric and U.S. customary measurement systems will be covered, students will not be converting between the two systems.

Students often confuse capacity and volume. The capacity of a container is the amount it can hold. Capacity is normally used for substances that can be poured: liquids, sugar, birdseed. The volume of the substance is the amount actually in the container. A 1-liter beaker may contain only a volume of 400 mL of water

Materials

- 1-gallon jugs
- 1-liter beakers or measuring cups
- 1-liter bottles
- 2-gallon buckets
- 2-liter bottles with tops cut off and labels removed
- Containers to hold liquid such as buckets/pails, drink bottles, jars, jugs, measuring cups, a large thermos, and a watering can
- Dry goods to measure such as sand, rice, or bird seed
- Large plastic drinking cups
- Large sponges
- Recording sheets
- Small paper cups
- Small pitchers (at least 2 different sizes) that students can easily handle when filled
- Water
- Whiteboards

Storybooks

- *Mathstart: Room For Ripley* by Stuart J. Murphy
- *Pastry School in Paris: An Adventure in Capacity* by Cindy Neuschwander
- *Capacity* by Henry Pluckrose

Chapter Opener

Objective
- Investigate capacity.

Lesson Materials
- Containers such as measuring cups, jugs, jars, drink bottles, or pails
- Large plastic drinking cups
- Small paper cups

Show students different containers such as those listed in **Lesson Materials**.

Have students predict which containers will hold the least and greatest amounts of water.

Provide pairs of students with a container, a large plastic drinking cup, and a small paper cup. Have them fill their containers two times, first using one of their cups and then using their other cup. They should keep track of how many of each size of cups it takes to fill their containers.

Students may see that if they use a larger cup it takes fewer cups than when using a smaller cup.

Just like when we compare length or weight, it helps to have a common unit when we measure how much a container holds.

Lesson 1 Comparing Capacity

Objectives

- Understand the meaning of capacity.
- Compare the volume of water in two or more containers by direct and indirect comparison.

Lesson Materials

- Small pitchers (at least 2 different sizes) that students can easily handle when filled
- Water or something dry like sand, rice, or bird seed
- Larger, identical containers (2-liter bottles with tops cut off and labels removed)
- Small cups of the same size

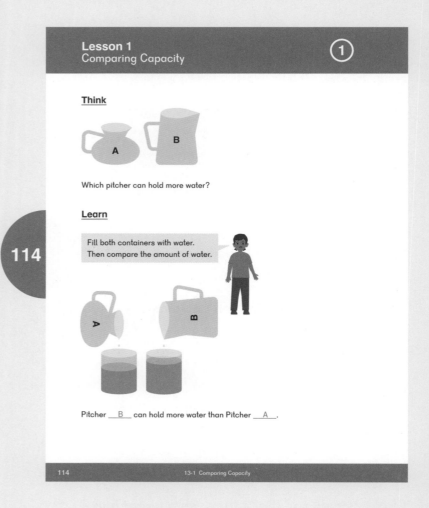

Think

Provide groups of students with 2 different shaped small pitchers filled to the top with water, 2 identical larger containers, and cups.

Pose the **Think** problem.

If this is not feasible in the classroom:

- Use rice instead of water to fill the pitchers.
- Have students work outside.

Ask students:

- Which pitcher holds more of the smaller cups? How do you know?
- Does a pitcher that is taller always hold more water? What about a pitcher that is rounder?

Learn

Discuss the examples with students. Ask how the textbook example is the same or different from the activity they did with their pitchers.

146   Teacher's Guide 2B Chapter 13   © 2017 Singapore Math Inc.

Alex compares by pouring the water from each pitcher into the same sized container to see that Pitcher B holds more water than Pitcher A.

Mei suggests measuring the amount of water in the pitchers using a small cup as a standard unit.

Define the term **capacity** as the maximum amount of something that a container can hold.

We say that Pitcher B has a greater capacity than Pitcher A because Pitcher B can hold more water.

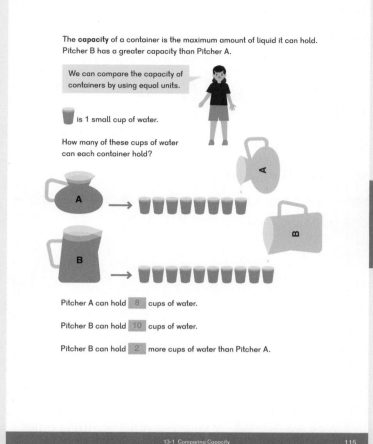

Do

① Without counting the cups, have students predict which container will hold the most water.

Activity

▲ **Capacity Relay**

Materials: For each team: recording sheet, large sponge, bucket of water, plastic cup, one-liter bottle, one-gallon jug (like gallon milk jug)

Divide the class into two teams.

Teams members take turns filling the cup by dipping the sponge into the bucket of water and squeezing the water out into the cup.

Once the cup is filled, a team member pours the cup into the drink bottle. The drink bottle won't be filled after a single cup, so the next team member races to fill up another cup. Once the drink bottle is filled, a team member records how many cups it took to fill the drink bottle.

When the drink bottle is full, the team players pour its contents into the plastic jug. The water from the drink container won't fill up the plastic jug. Team members will have to race to fill up a cup again and figure out how many cups will fill up this plastic jug.

The winner is the first team to fill up their plastic jug and have the correct answers written down (how many cups are in each container).

Exercise 1 • page 129

Lesson 2 Units of Capacity

Objective

- Measure capacity in liters, quarts, and gallons.

Lesson Materials

- 1-liter beaker or measuring cups
- 2-gallon bucket
- Small paper cups, similar in size
- Various containers (See ❸ on textbook page 122)

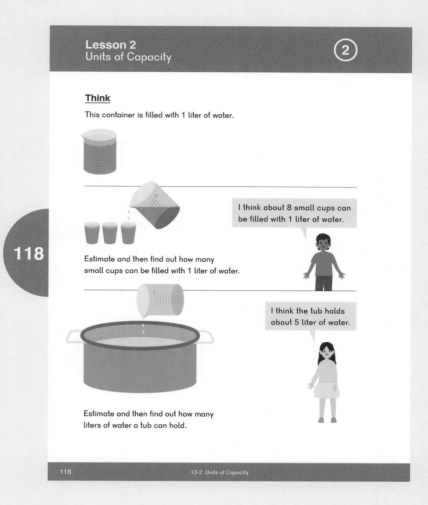

Think

Divide the class into two groups. Have each group estimate how many cups of water can be filled with 1 liter of water, and how many liters a bucket of water will hold.

Give groups 1-liter beakers of water. Have each group find and record how many of the paper cups one liter of water will fill.

Repeat the activity with the 2-gallon bucket.

Have groups share their estimates and answers. Compare the answers between the groups. Most likely, the two groups have found that the number of cups needed to empty the 1-liter beaker differs from the number of cups needed to empty the bucket.

Discuss why the groups got different numbers.

Just as when students measured length, the numerical value of what we measure depends on the unit we use. To compare, we need to use something that measures the same for everyone.

Do ❸ on page 122 while water and containers are out. Ensure students have a feeling for how much a liter is by using containers from Lesson 1. Have them estimate, then find the number of liters a container will hold.

Learn

Note that the **Dimensions Math**® series will use a capital L to denote liters.

Ask students why the terms "almost" and "about" are used when filling containers. Remind students that measurement is always an estimate. Additionally, students should notice that it is difficult to fill the containers completely to the top.

Do

1. Students should recognize that a taller container does not necessarily mean that the container holds more liquid.

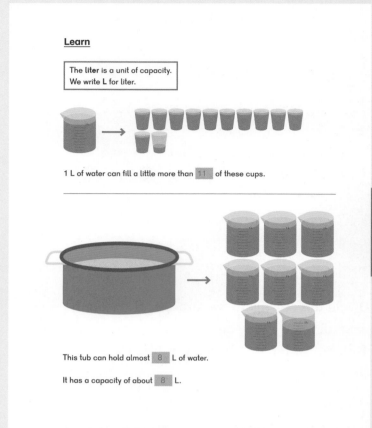

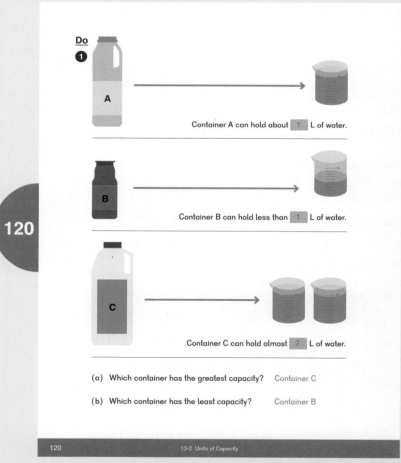

③ Students should have completed this task during the **Think** portion of the lesson.

④ This problem simply introduces quarts and gallons as different units of measurement.

Quarts are combined to make a larger unit of measurement, the gallon.

Students should consider that 1 gallon is 4 quarts:

2 gallons is 4 quarts × 2, or 8 quarts

3 gallons is 4 quarts × 3, or 12 quarts

4 gallons is 4 quarts × 4, or 16 quarts...

Ask students what other units of measuring liquid they know.

Ask students to name things that come in 1 gallon containers.

Students should recall that they measured length in two systems, metric (centimeters and meters), and U.S. customary (inches and feet). Capacity can also be measured in different units. Liters are a metric measurement. Quarts and gallons are a U.S. customary measurement of capacity.

Exercise 2 • page 133

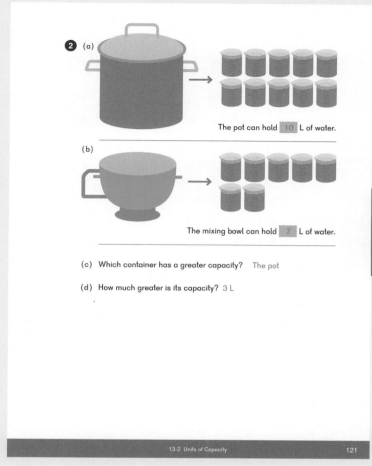

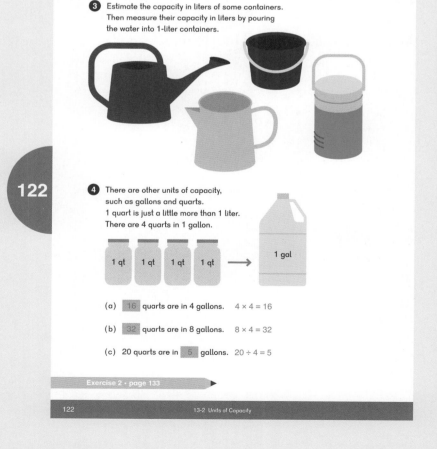

Lesson 3 Practice

Objective

- Practice measuring and comparing capacity.

After students complete the **Practice** in the textbook, have them continue to practice concepts with activities from the chapter.

6 Ensure that students understand that there is a standard unit of measurement called a "cup," as opposed to tea cups or coffee cups that can hold different amounts.

Exercise 3 • page 137

Brain Works

★ Buckets

Alice has a large thermos she wishes to fill with 14 liters of water. She only has 1 container marked 5 liters and 1 container marked 3 liters to use to fill the thermos. If she can fill her containers multiple times, how could Alice fill the thermos exactly without wasting any water?

Fill the 5-liter container once and the 3-liter container 3 times.

Lesson 3 Practice — P 3

1. An art teacher bought 12 gallons of paint. After the students did a painting project, she had 4 gallons of paint left. How much paint did they use for the project?
   12 − 4 = 8;
   8 gallons

2. Fish tank A has a capacity of 62 L of water. Fish tank B has a capacity of 56 L of water.
   (a) Which tank can hold more water? Tank A
   (b) How much more? 62 − 56 = 6
   6 L

3. The table shows the amount of water each child drank in 2 weeks.

Amelia	15 L
Evan	23 L
Hailey	18 L

   (a) Who drank the greatest amount of water? Evan
   (b) Who drank the least amount of water? Amelia
   (c) How much more water did Evan drink than Amelia? 23 − 15 = 8; 8 L
   (d) How much less water did Hailey drink than Evan? 23 − 18 = 5; 5 L

4. A bucket can hold 10 L of water. Isaac poured 5 buckets of water in an empty tank. How much water is in the tank?
   10 × 5 = 50;
   50 L

5. Violet buys 24 L of bottled water. The water comes in jugs that hold 3 L of water. How many jugs does she buy?
   24 ÷ 3 = 8;
   8 jugs

6. Cups are another unit of measure, often used in cooking. There are 4 cups in 1 quart.
   (a) Malik bought 3 quarts of cooking oil. How many cups of oil did he buy?
   3 × 4 = 12; 12 cups
   (b) How many 1-quart jars are needed for 8 cups of jam?
   8 ÷ 4 = 2; 2 quart jars

7. 1 L of water was poured into each of these containers. They each have the same amount of water. Explain why the level of water is different in each.
   The water level will be lower when the container is wider.

Exercise 3 • page 137

Exercise 1 • pages 129–132

Chapter 13 Capacity

Exercise 1

Basics

1. The diagram shows how much water each container can hold.

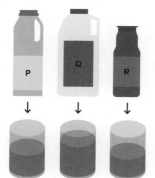

Fill in the blanks.

(a) Container P can hold less water than Container __Q__ and more water than Container __R__.

(b) Container __Q__ has the greatest capacity.

(c) Container __R__ has the least capacity.

2. ▭ is 1 unit.
The picture shows how many units of water will fill the bowls.

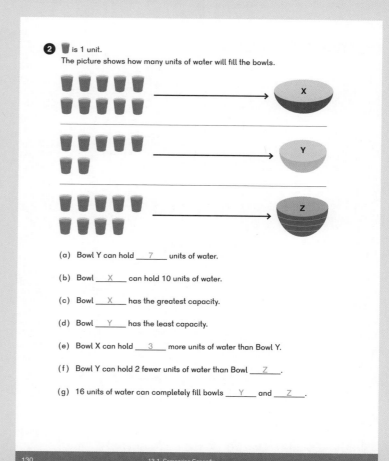

(a) Bowl Y can hold __7__ units of water.

(b) Bowl __X__ can hold 10 units of water.

(c) Bowl __X__ has the greatest capacity.

(d) Bowl __Y__ has the least capacity.

(e) Bowl X can hold __3__ more units of water than Bowl Y.

(f) Bowl Y can hold 2 fewer units of water than Bowl __Z__.

(g) 16 units of water can completely fill bowls __Y__ and __Z__.

Practice

3. ▭ is 1 unit.

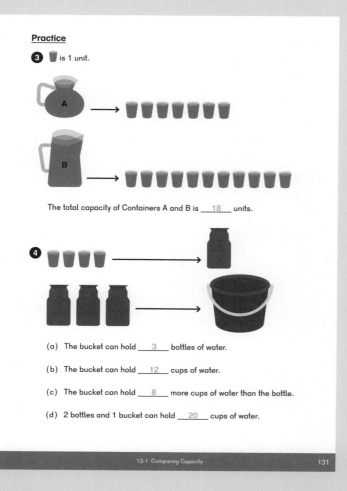

The total capacity of Containers A and B is __18__ units.

4.

(a) The bucket can hold __3__ bottles of water.

(b) The bucket can hold __12__ cups of water.

(c) The bucket can hold __8__ more cups of water than the bottle.

(d) 2 bottles and 1 bucket can hold __20__ cups of water.

Challenge

5.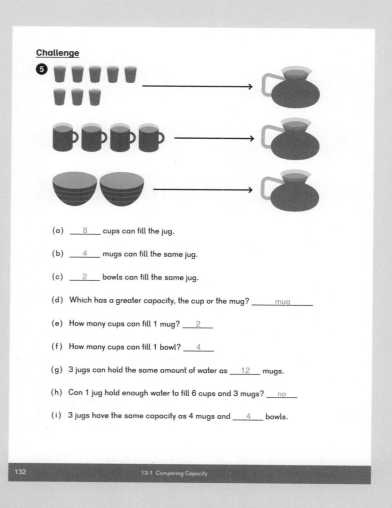

(a) __8__ cups can fill the jug.

(b) __4__ mugs can fill the same jug.

(c) __2__ bowls can fill the same jug.

(d) Which has a greater capacity, the cup or the mug? __mug__

(e) How many cups can fill 1 mug? __2__

(f) How many cups can fill 1 bowl? __4__

(g) 3 jugs can hold the same amount of water as __12__ mugs.

(h) Can 1 jug hold enough water to fill 6 cups and 3 mugs? __no__

(i) 3 jugs have the same capacity as 4 mugs and __4__ bowls.

Exercise 2 • pages 133–136

Exercise 2

Basics

1 Look for some containers you think hold about 1 liter.
Use a 1-liter container to measure their capacity.
Put a check (✓) in the correct box.

Answers will vary.

Container	Less than 1 L	1 L	More than 1 L

2 Look for some containers that hold more than 1 liter.
Estimate about how many liters of water they hold.
Then use the 1-liter container to find their capacity.

Answers will vary.

Container	Estimated	Measured
	About _____ L	Between _____ L and _____ L
	About _____ L	Between _____ L and _____ L
	About _____ L	Between _____ L and _____ L
	About _____ L	Between _____ L and _____ L

13-2 Units of Capacity   133

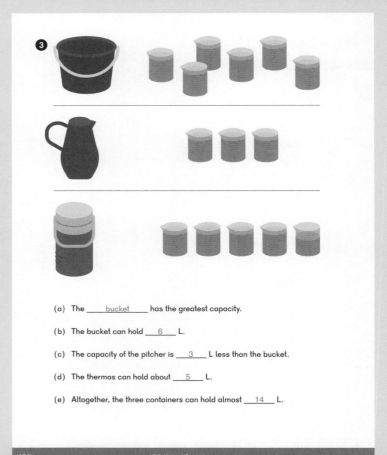

3

(a) The ___bucket___ has the greatest capacity.

(b) The bucket can hold ___6___ L.

(c) The capacity of the pitcher is ___3___ L less than the bucket.

(d) The thermos can hold about ___5___ L.

(e) Altogether, the three containers can hold almost ___14___ L.

Practice

4 Circle the most reasonable capacity for each of the following.

(a) A fish tank in the home.

  1 L   (40 L)   800 L

(b) A bottle of juice.

  (1 L)   10 L   50 L

(c) A bath tub.

  20 L   (200 L)   900 L

5 A water tank can hold 400 L of water.
It has 132 L of water in it now.
How much more water is needed to fill the tank?

400 − 132 = 268
268 L is needed to fill the tank.

6 A pot has 36 L of broth.
Ms. Ivanov ladled 4 L of broth into each jar.
How many jars did she use?

36 ÷ 4 = 9
She used 9 jars.

Challenge

7 Each water bottle can hold 1 L of water.
How many liters can the bucket hold?
1 + 3 + 5 = 9 L
The bucket can hold 9 L.

8 Using only these three containers, how can you divide the water in the 12-L container evenly between all 3 containers?

Pour from the 12-L container to fill the 4-L container.
Pour from the 4-L container into the 9-L container.
Pour another 4 L from the 12-L container into the 4-L container.
They now all have 4 L of water.

Teacher's Guide 2B Chapter 13

Exercise 3 • pages 137–140

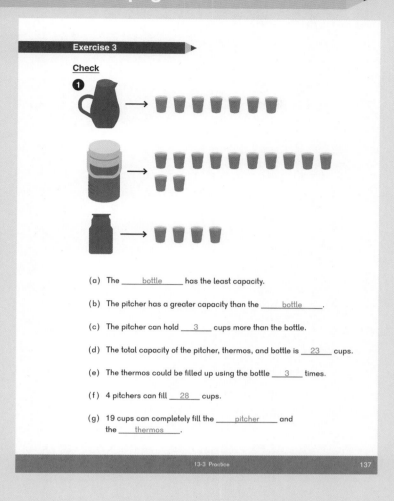

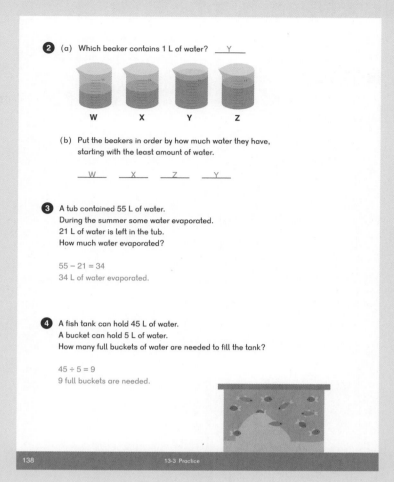

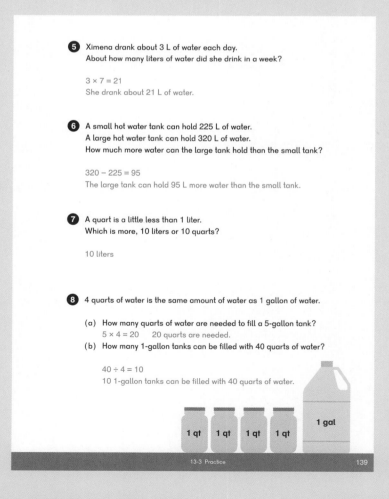

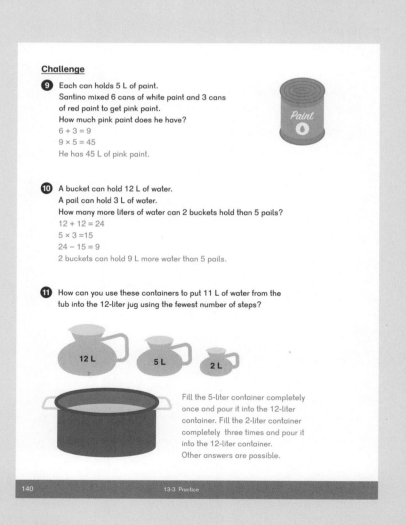

Notes

Chapter 14 Graphs — Overview

Suggested number of class periods: 3–4

	Lesson	Page	Resources	Objectives
	Chapter Opener	p. 159	TB: p. 125	Collect data. Investigate graphs.
1	Picture Graphs	p. 160	TB: p. 126 WB: p. 141	Interpret a scaled picture graph.
2	Bar Graphs	p. 163	TB: p. 130 WB: p. 145	Interpret bar graphs with a scale of 1 unit.
3	Practice	p. 165	TB: p. 134 WB: p. 149	Practice reading and interpreting graphs.
	Workbook Solutions	p. 167		

Chapter 14 Graphs

Notes & Materials

In **Dimensions Math® 1B** Chapter 11: Comparing, students learned to:

- Read and interpret picture graphs based on a one-to-one representation.
- Compare quantities represented on a picture graph.

In this chapter, students will learn to:

- Create a graph from tally marks.
- Create and read a scaled picture graph.
- Create and read a bar graph with one-to-one representation.
- Understand what kind of information can be determined from the graph.

In Lesson 1: Picture Graphs, students will read picture graphs where each symbol or picture will represent either 2, 3, 4, 5, or 10 of the surveyed data. This reinforces the multiplication facts learned in previous **Dimensions Math® 2A** and **2B** chapters.

Graphs are pictorial representations of data that are helpful in making comparisons between sets of objects. It is easy to see the difference on a graph between quantities of objects.

Scaled graphs are used when the quantities or the range of data are large.

Students will **not** be using partial symbols to represent quantities, for example:

 Stands for 10 students

 Stands for 5 students

Therefore, any data students use to create their own scaled picture graphs will have to be provided so that the quantities are multiples of 2, 3, 4, or 5.

In Lesson 2: Bar Graphs, students will transition from a scaled picture graph to a bar graph with a numerical scale on the side. At this level, the scale is only 1, that is, the interval is only 1. In **Dimensions Math® 3A** Chapter 7: Graphs and Tables, students will work with scaled bar graphs.

Materials

- Crayons or markers
- Recording sheets
- Sticky notes
- Whiteboards

Blackline Masters

- Bar Graph 14-2
- Math Mazes

Chapter Opener

Objectives
- Collect data.
- Investigate graphs.

Lesson Materials
- Sticky notes

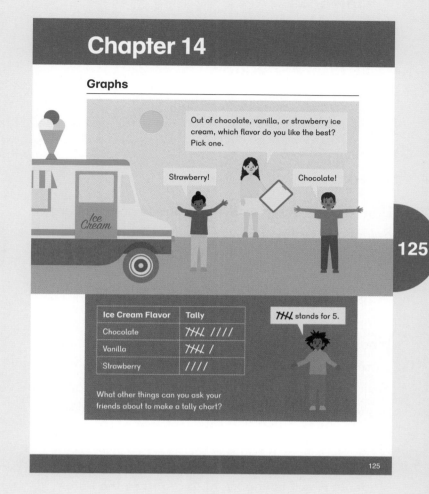

Have students discuss textbook page 125.

Have students work in small groups to come up with a question and survey their classmates. Limit students to 4 or 5 categories.

Ideas:

- Of the choices apple, melon, banana, and peach, which fruit do you like best?
- What type of pet do you have? Dog, cat, fish, none, or other?
- How many pets do you have?
- How many people are in your family?
- How do you get to school? Bus, car, bike, or walk?

Have students record the results with tally marks and share with the class.

As review, choose one of the surveys and create a graph on the board from the data, using sticky notes for each tally mark.

Alternatively, groups can create a basic graph from their data.

In the next lesson, students will create scaled graphs. Depending on the quantities in each category, student groups could use the same data to create a scaled graph after Lesson 1 and a Bar Graph after Lesson 2. At the end of the chapter, groups should have 4 different representations of the data collected in the **Chapter Opener**:

- Tally marks
- Picture graph
- Scaled picture graph
- Bar graph

Activity

▲ Math Facts

Students will be creating and interpreting scaled graphs. Prepare by reviewing multiplication facts for 2, 3, 4, 5, and 10 by playing games from **Dimensions Math® 2A** Chapter 7: Multiplication and Division of 2, 5, and 10 and **2B** Chapter 9: Multiplication and Division of 3 and 4.

Lesson 1 Picture Graphs

Objective

- Interpret a scaled picture graph.

Lesson Materials

- Sticky note for each student

Think

Draw a large graph with 4 categories on the board. Select one of the surveys from the **Chapter Opener** to graph.

Have students put a sticky note in the category that represents their choice from the survey.

When all students have placed their sticky note in a category, check to see if the categories have an even number of notes. If not, add notes for yourself or others in the school to ensure that each category has a multiple of 2.

The notes will be haphazardly placed. Discuss with students why it is important to organize the notes to make the graph easy to read. Line up the sticky notes similar to the example at right, and add a key at the bottom of the graph as shown.

Ask students questions such as:

- What can you tell from the graph?
- How many sticky notes are in the ____ column?
- How many more students wore ____ than ____?

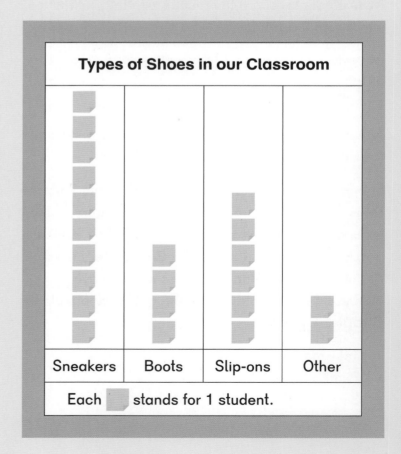

160   Teacher's Guide 2B Chapter 14   © 2017 Singapore Math Inc.

Change the totals in the columns and the key at the bottom of the graph:

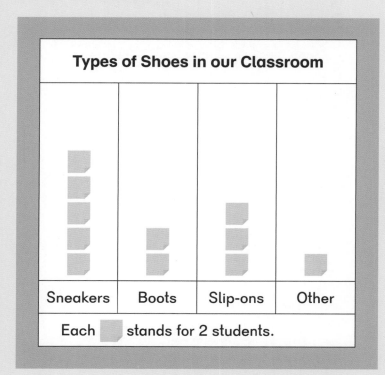

Ask students, "How does changing the key below change the information in the graph? How many does each category have now?"

Compare Alex's graph with the one created in **Think**.

Learn

As students answer the questions, ask:

- How many of each item were sold?
- Why are Emma and Dion thinking about multiplication equations?

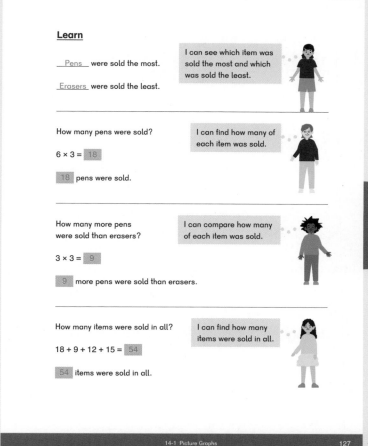

Do

1 Ask students to come up with their own questions. Have them speculate why there is higher attendance on Friday. One reason people create surveys and graphs to help find information.

Ask students:

- Why might a museum graph attendance at an art show?
- Why might it be important to know which day is busiest?

2 Have students come up with their own questions. Ask, "Why might the friends want to graph the amount of money they have saved?" (To see who is closest to a saving goal, to graph a competition, etc.)

Activity

▲ Scaled Picture Graph

Materials: Data from the **Chapter Opener** in tally marks, recording sheet, crayons or markers

Have student groups create a scaled picture graph from the data they collected from the **Chapter Opener**.

Have students choose the best scale to represent their data.

Not all data will scale (a tally of 7 can't be scaled perfectly by 2, 3, 4, or 5). Rather than arbitrarily change the data, assign quantities and scales for students to use for their graphs.

Exercise 1 • page 141

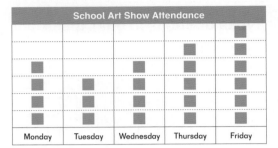

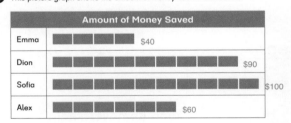

162   Teacher's Guide 2B Chapter 14   © 2017 Singapore Math Inc.

Lesson 2 Bar Graphs

Objective

- Interpret bar graphs with a scale of 1 unit.

Lesson Materials

- Bar Graph 14-2 (BLM)

Think

Create a bar graph on the board similar to the one in **Think**, or project the one on textbook page 130.

Ask students to discuss how this graph is different from the picture graphs they have worked with so far.

They may note that:

- There are numbers going up the side.
- The whole box is colored in.
- They can tell how many students chose each subject by looking at the greatest number the bars go up to.

Learn

Have students read the bar graph and discuss their answers.

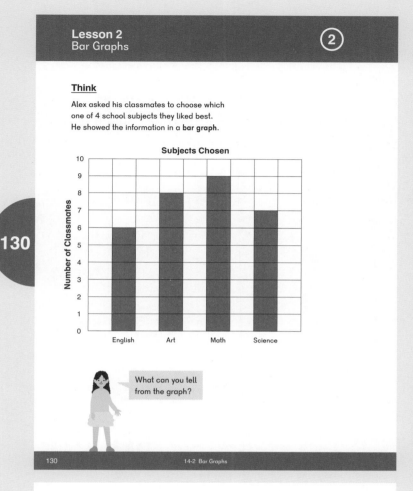

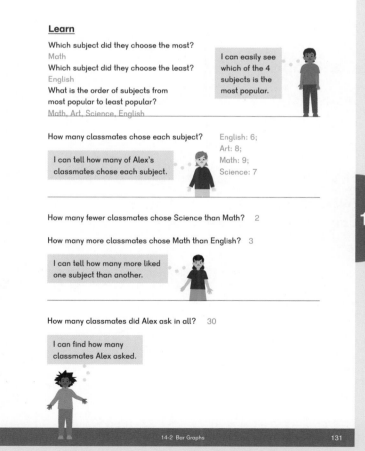

© 2017 Singapore Math Inc. Teacher's Guide 2B Chapter 14 163

Do

1 (g) and (h) Students should see that when collecting data it may be organized naturally; however, when presenting data, it is often appropriate to rearrange the columns to make it easier to read. Students should understand this occurs for presentation, and does not change the data.

2 Use Bar Graph 14-2 (BLM) as a template.

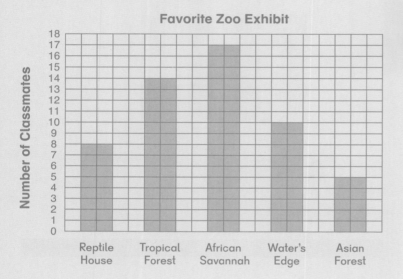

Activity

▲ Bar Graph

Materials: Data from the **Chapter Opener** in tally marks, Bar Graph 14-2 (BLM), recording sheet, crayons or markers

Have student groups create a bar graph from the data they collected from the **Chapter Opener**. They can use Bar Graph 14-2 (BLM).

Exercise 2 · page 145

Do

1 This bar graph shows the number of 4 types of raptors rescued and released by a wildlife rescue society last year.

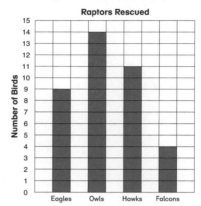

(a) Which type of bird had the greatest number rescued? Owls
(b) Which type of bird had the least number rescued? Falcons
(c) How many more eagles were rescued than falcons? 5
(d) How many of each type of bird were rescued? Eagles: 9; Owls: 14; Hawks: 11; Falcons: 4
(e) Turkey vultures were also rescued, but their information is not shown on the graph. Twice as many falcons were rescued as turkey vultures. How many turkey vultures were rescued? 2
(f) How many birds, including turkey vultures, were rescued in all? 40

(g) This graph shows the same information. What is different about how the information is organized?

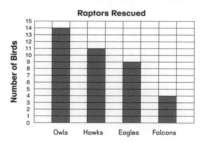

(h) What advantage is there to organizing this information this way?
Possible answer: It is easier to compare most to least.

2 Mei asked her classmates which zoo exhibit they liked best and recorded the information with tally marks.

Exhibit	Tally
Reptile House	𝍷𝍷𝍷𝍷𝍷 ///
Tropical Forest	𝍷𝍷𝍷𝍷𝍷 𝍷𝍷𝍷𝍷𝍷 ////
African Savannah	𝍷𝍷𝍷𝍷𝍷 𝍷𝍷𝍷𝍷𝍷 𝍷𝍷𝍷𝍷𝍷 //
Water's Edge	𝍷𝍷𝍷𝍷𝍷 𝍷𝍷𝍷𝍷𝍷
Asian Forest	𝍷𝍷𝍷𝍷𝍷

Make a bar graph using this information.

Exercise 2 · page 145

Lesson 3 Practice

Objective

- Practice reading and interpreting graphs.

After students complete the **Practice** in the textbook, have them continue to practice reading graphs with activities from the chapter or materials from other school subjects.

Ask students, "Which of these graphs are picture graphs and which are bar graphs?"

❸ Ask students if they notice anything different about this graph. Ask them what it means when a category is blank.

134

Lesson 3 Practice — P 3

1 Some children helped clean up a park.
This picture graph shows how many empty plastic bottles each child found and put in recycling.

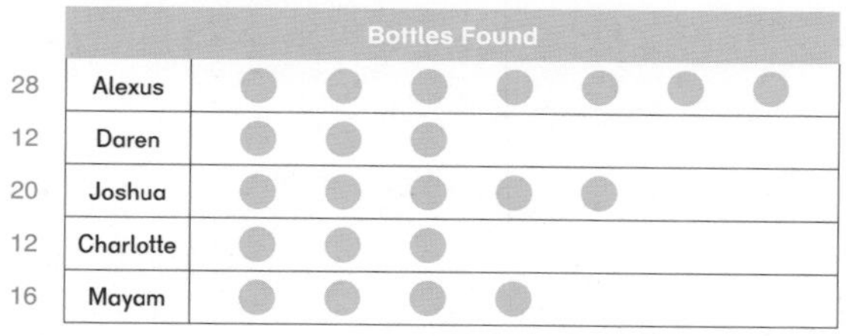

Bottles Found

28	Alexus	●	●	●	●	●	●	●
12	Daren	●	●	●				
20	Joshua	●	●	●	●	●		
12	Charlotte	●	●	●				
16	Mayam	●	●	●	●			

Each ● stands for 4 bottles.

(a) Who found the most bottles? Alexus
(b) Which two children found same number of bottles? Daren and Charlotte
(c) How many more bottles did Alexus find than Mayam? 12
(d) How many fewer bottles did Daren find than Joshua? 16
(e) How many bottles did each of these 5 children find? See chart
(f) Elena found 24 bottles.
How many ● would be used to add her information to the graph? 6 circles
(g) How many bottles did all 6 children, including Elena, find? 112

135

2 Brianna sold drinks at the Track and Field Day.
This picture graph shows how many of each kind she sold.

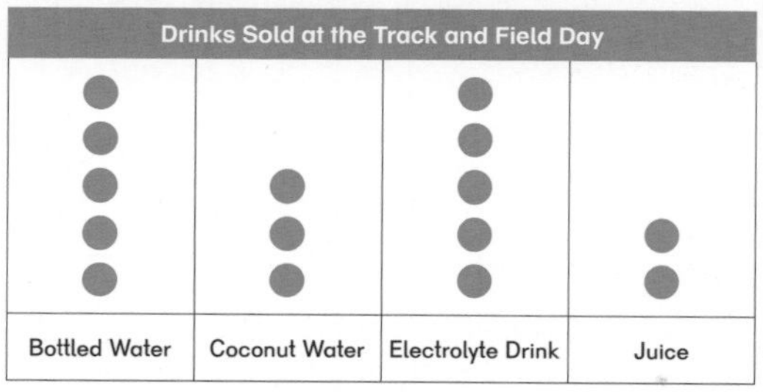

Drinks Sold at the Track and Field Day

Bottled Water	Coconut Water	Electrolyte Drink	Juice
●		●	
●	●	●	
●	●	●	●
●	●	●	●
●	●	●	

Each ● stands for 2 drinks.

(a) Use the information in the graph to complete the table.

Number of Type of Drink Sold

Bottled Water	Coconut Water	Electrolyte Drink	Juice
10	6	10	4

(b) Which type of drink was the least popular? Juice
(c) How many Bottled Waters and Coconut Waters did Brianna sell? 16
(d) How many fewer Juices did she sell than Bottled Waters? 6
(e) She sold the Electrolyte Drinks for $5 each. $5 \times 10 = 50$;
How much money did she receive for the Electrolyte Drinks? $50
(f) She received $20 from selling the Bottled Waters.
What was the cost of 1 Bottled Water? $20 \div 10 = 2$; $2
(g) How can this type of graph help Brianna decide what type of drinks to bring to the next Track and Field Day?
Possible answer: Bottled Water and Electrolytes Drink were most popular, so bring more of them. Juice was least popular, find out why. Too expensive?

Exercise 3 • page 149

Brain Works

★ Math Maze

Materials: Math Mazes (BLM)

Begin the maze with zero.

Follow the arrows to find the total listed. There are two possible paths through each maze.

For extra challenge, find that path that ends with the highest number and the lowest number.

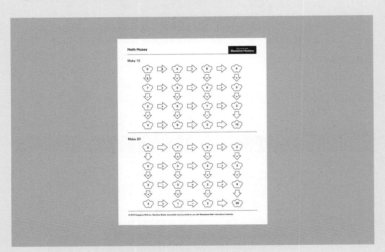

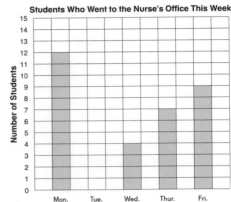

Exercise 1 • pages 141–144

Chapter 14 Graphs

Exercise 1

Basics

1. Members of a club voted to choose the color for their new club t-shirt. This picture graph shows the results.

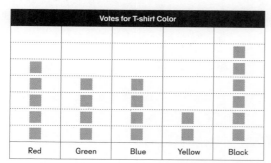

Each ▦ stands for 5 votes.

(a) ___Black___ got the most votes.

(b) ___10___ club members voted for yellow.

(c) ___15___ more club members voted for red than for yellow.

(d) ___10___ fewer club members voted for blue than for black.

(e) The colors ___green___ and ___blue___ received the same number of votes.

Practice

2. Count the number of each kind of bug.
   Then color the circles in to complete the graph on the next page.

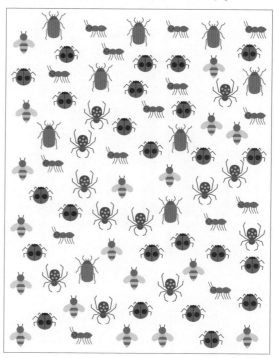

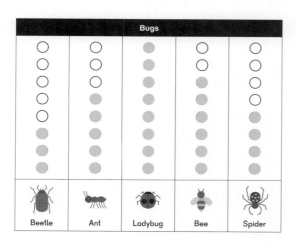

Each ● stands for 3 bugs.

(a) There are ___24___ ladybugs.

(b) There are ___3___ more spiders than beetles.

(c) There are ___9___ fewer ants than ladybugs.

(d) There are ___18___ more ladybugs and bees than beetles and ants.

(e) Put the bugs in order of least to greatest number counted.

   ___Beetle___, ___Spider___, ___Ant___, ___Bee___, ___Ladybug___

3. This picture graph shows which fruit some children chose for snack.

Fruit Chosen							
Apple	☺	☺	☺	☺			
Grapes	☺	☺	☺	☺	☺	☺	☺
Banana	☺	☺	☺	☺	☺	☺	
Pear	☺	☺	☺				

Each ☺ stands for 2 children.

(a) The children chose ___grapes___ the most.

(b) ___14___ children chose bananas.

(c) ___4___ fewer children chose pears than apples.

(d) The graph shows the choices of ___50___ children.

4. Each ◆ stands for 4 cars.

(a) ◆◆◆◆ stands for ___16___ cars.

(b) Color the correct number of shapes to show 24 cars.

   ◆◆◆◆◆◆◇◇◇

5. ▲▲▲▲▲ stands for 50 plants.

    stands for ___20___ plants.

Exercise 2 • pages 145–148

Exercise 2

Basics

① This bar graph shows the number of times certain pizza toppings were ordered one day at a pizzeria.

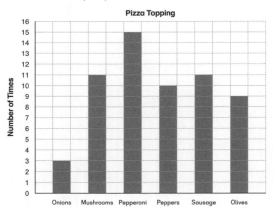

(a) __Pepperoni__ was ordered the most.

(b) __Onions__ were ordered the least.

(c) Olives were ordered __9__ times.

(d) Peppers were ordered __5__ fewer times than pepperoni.

(e) __Mushrooms__ and __sausage__ were ordered the same number of times.

Practice

② Count the number of each kind of sticker (circle, moon, star, square, and triangle).
Complete the bar graph.

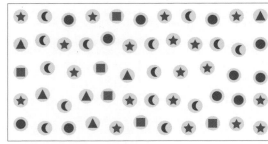

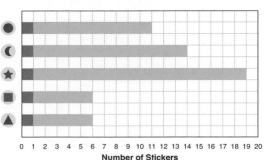

③ Complete the bar graph below using the information from this picture graph.

Each 😊 stands for 2 children.

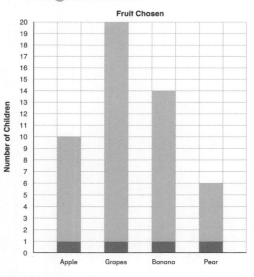

Challenge

④ Use the information below to complete the bar graph.

Riya has 13 stamps.
Taylor has 9 more stamps than Riya.
Pedro has 6 fewer stamps than Taylor.
Liam has the same number of stamps as Pedro.
Altogether, the five children have 80 stamps.

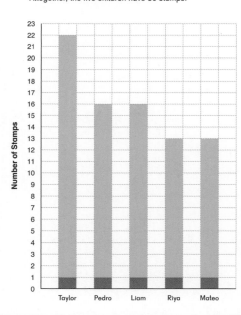

Exercise 3 • pages 149–152

Exercise 3

Check

1. This table shows the number of bottles collected by some children for a recycling project.

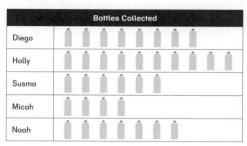

Each 🍶 stands for 3 bottles.

(a) _____Holly_____ collected the most bottles.

(b) Diego collected __24__ bottles.

(c) Holly collected __9__ more bottles than Noah.

(d) Susma collected 6 fewer bottles than _____Diego_____.

(e) Micah collected more bottles than is recorded on the graph.
He collected a total of 24 bottles.
How many more 🍶 should be drawn on the graph for Micah's bottles?

4 more

2. This table shows the number of four types of fish sold at a fish store.

Fish	Goldfish	Swordtail	Guppy	Angelfish
Number	40	35	25	15

(a) Use this information to complete the picture graph below.

(Fish Sold picture graph with columns: Goldfish, Swordtail, Guppy, Angelfish, Clownfish)

Each 🐟 stands for 5 fish.

(b) 15 fewer clownfish were sold then swordtail.
Draw 🐟 for the number of clownfish sold.

(c) How many of the 5 types of fish were sold in all? 135

(d) If the angelfish and clownfish cost $3 each, how much more money was collected from selling clownfish than angelfish? $15

3. This table is a tally for the colors of some cars in a parking lot.

Red									
Gray													
Black											
White									

(a) Use the information to complete the bar graph.

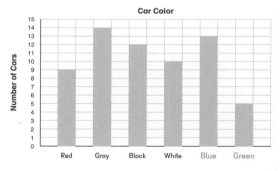

(b) There were 4 more blue cars than red cars and 8 fewer green cars than blue cars.
Add this information to the graph.

(c) __63__ cars were counted altogether.

(d) Put the car colors in order from greatest to least number counted.

__Gray__, __Blue__, __Black__, __White__, __Red__, __Green__

Challenge

4. This picture graph shows the number of burritos sold by some children at a fund raiser last week.
Use the information below to write the correct name in each column.

Matias sold 32 burritos.
Paula sold 20 burritos fewer than Matias.
Misha sold 4 more burritos than Paula.
Nolan sold twice as many burritos as Paula.

Each burrito cost $3.
Sasha raised $24 selling burritos.
Complete the graph for the number of burritos Sasha sold.

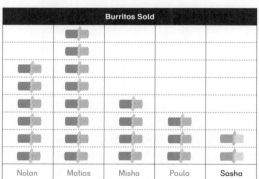

Each 🌯 stands for 4 burritos.

Notes

Chapter 15 Shapes — Overview

Suggested number of class periods: 8–9

Lesson		Page	Resources	Objectives
	Chapter Opener	p. 175	TB: p. 137	Investigate shapes and patterns.
1	Straight and Curved Sides	p. 176	TB: p. 138 WB: p. 153	Draw straight lines and curved lines. Identify open and closed shapes.
2	Polygons	p. 178	TB: p. 142 WB: p. 155	Identify, describe, and categorize polygons by attributes.
3	Semicircles and Quarter-circles	p. 182	TB: p. 147 WB: p. 159	Identify semicircles and quarter-circles. Make new shapes by combining semicircles and quarter-circles.
4	Patterns	p. 184	TB: p. 150 WB: p. 163	Make and complete patterns with two-dimensional shapes according to one or two attributes and explain the patterns.
5	Solid Shapes	p. 187	TB: p. 154 WB: p. 167	Identify attributes of three-dimensional shapes.
6	Practice	p. 190	TB: p. 159 WB: p. 171	Practice concepts from the chapter.
	Review 4	p. 192	TB: p. 161 WB: p. 175	Review content from Chapter 1 through Chapter 15.
	Review 5	p. 194	TB: p. 165 WB: p. 181	Review content from Chapter 1 through Chapter 15.
	Workbook Solutions	p. 196		

Chapter 15 Shapes   Notes

In **Dimensions Math® KA**, students learned to:

- Identify and name three-dimensional shapes: cube, sphere, cylinder, and cone.
- Identify and name two-dimensional shapes: square, circle, rectangle, triangle.

In **Dimensions Math® 1A** Chapter 8: Shapes, students learned to:

- Identify the four basic shapes (circle, rectangle, square, and triangle) on three-dimensional shapes.
- Group shapes according to size, shape, color, and orientation.
- Create and identify patterns.
- Compose shapes out of other shapes.

In this chapter, students will:

- Investigate straight and curved lines.
- Describe and classify two-dimensional shapes based on the number of sides and corners (or angles, although the term "angle" will not be used).
- Identify polygons and specific polygons (triangles, quadrilaterals, pentagons, and hexagons).
- Create and identify patterns based on attributes.
- Describe and classify solid or three-dimensional shapes by faces, edges, and corners.

Attributes that define two-dimensional shapes are sides and corners (vertices). Attributes that define three-dimensional shapes are faces, edges, and corners (vertices — straight or curved sides, relative lengths of the sides, sizes of the angles where the sides meet).

Shapes are not defined by orientation, color, or size. Students often will say that a rotated square is a diamond. However, patterns based on shapes are defined by shape, orientation, color, and size.

Manipulating shapes by flipping, rotating, and sliding them helps develop visual and spatial awareness and lays the foundation for later geometry concepts. Students should be allowed plenty of time to investigate with shapes.

This chapter includes many ideas and terminology. Students may already be familiar with many of the topics, as there are shapes and patterns in the world around them that they can observe every day.

Geometry terminology

Attribute: A characteristic

Closed shape: A figure that can be traced with the same starting and stopping points, and without crossing or retracing any part of the figure.

Corner: Due to the abundance of vocabulary in the chapter, the term "corner" will be used instead of "angle" or "vertex."

Cube: All faces are squares

Cuboid: Also known as a rectangular prism

Edge: Where two sides meet

Face: A surface on a three-dimensional object

Hexagon: A polygon with 6 sides

Orientation: The position or direction of a shape compared with the other shapes

Pentagon: A polygon with 5 sides

Polygon: A closed shape with straight sides

Quadrilateral: A polygon with 4 sides

Quarter-circles: One-fourth of a circle

Rotate: Turn an object or shape

Semicircle: A half circle

Sides: Lines that join corners (or vertices) in a polygon.

Straight and curved lines: It is important that students distinguish between straight and curved lines as in later geometry, a line is not something we draw with a pencil, but a mathematical object with a specific definition (two points define an infinite line in space).

Chapter 15 Shapes

Materials

- Crayons or colored pencils
- Geoboards
- Geometric solids shape sets
- Large paper shapes including quadrilaterals, rectangles, circles, semicircles, quarter-circles, triangles, pentagons, and hexagons
- Music
- Painter's tape
- Paper
- Paper clip
- Paper squares
- Pattern blocks
- Round paper plates or paper circles
- Rubber bands
- Ruler
- Scissors
- Shape cards including quadrilaterals, rectangles, circles, semicircles, quarter-circles, triangles, pentagons, and hexagons
- Tangram shapes
- Three-dimensional solids such as canisters, boxes, soccer balls, party hats, dice, and frisbees
- Two-color counters
- Whiteboards

Blackline Masters

- Connect 4 Solids
- Dot Paper
- Pencil-tracing Puzzles
- Semicircles

Storybooks

- *The Shape Of Things* by Dayle Ann Dodds
- *Shapes, Shapes, Shapes* by Tana Hoban
- *Cubes, Cones, Cylinders and Spheres* by Tana Hoban
- *Shape by Shape* by Suse MacDonald
- *Grandfather Tang's Story* by Ann Tompert
- *Mouse Shapes* by Ellen Stoll Walsh
- *Skippyjon Jones: Shape Up* by Judy Schachner
- *I Spy Shapes in Art* by Lucy Micklethwait

Notes

Chapter Opener

Objective
- Investigate shapes and patterns.

Lesson Materials
- Pattern blocks

Have students make pictures with pattern blocks.

Have them share their pictures. While students share, review names and attributes of shapes that students have already learned:

- Triangle
- Rectangle
- Square
- Circle

Lesson 1 Straight and Curved Sides

Objectives

- Draw straight lines and curved lines.
- Identify open and closed shapes.

Lesson Materials

- Paper
- Scissors
- Three-dimensional solids such as canisters, boxes, and frisbees

Think

Provide students with sheets of paper and have them draw shapes with straight and curved sides. Students can trace the bottoms of three-dimensional shapes.

Students should cut the shapes out and put them together to make other shapes or pictures.

Learn

Have students discuss the different lines and shapes.

Students do not need to learn the formal geometric definition of a line, but rather to recognize curved and straight lines.

Discuss the difference between open and closed shapes.

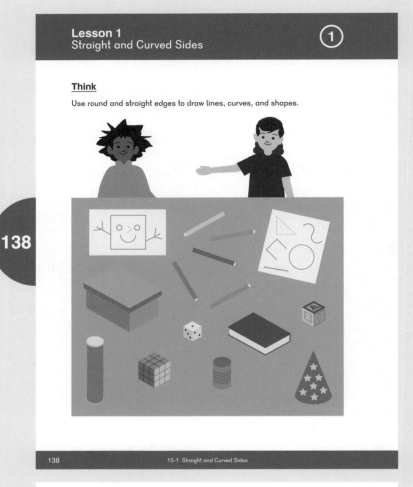

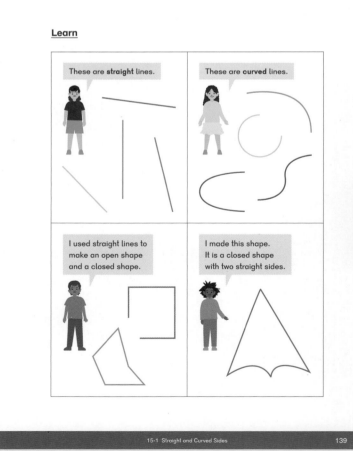

176   Teacher's Guide 2B Chapter 15   © 2017 Singapore Math Inc.

Do

Compare open shapes and closed shapes. Have students look at other shapes to determine if they are open or closed. For example, is this a shape, and is it open or closed?

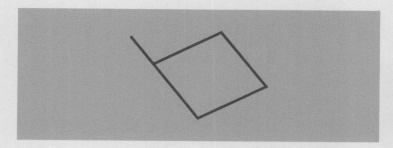

It is not considered to be closed, since it has an intersection and a side that does not connect.

Activity

▲ **Pencil-tracing Puzzles**

Materials: Pencil-tracing Puzzles (BLM)

Can you trace the puzzles on Pencil-tracing Puzzles (BLM) without lifting your pencil or tracing the same line segment more than once?

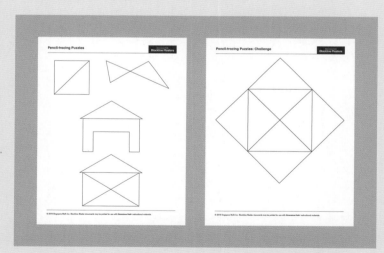

Exercise 1 • page 153

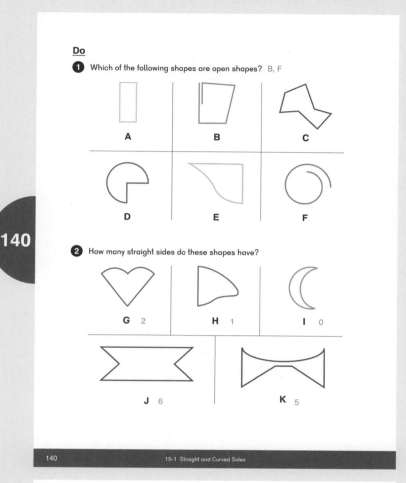

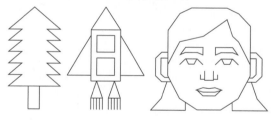

Lesson 2 Polygons

Objective

- Identify, describe, and categorize polygons by attributes.

Lesson Materials

- Geoboards
- Rubber bands
- Dot Paper (BLM)
- Ruler (to draw straight lines)
- Pattern blocks

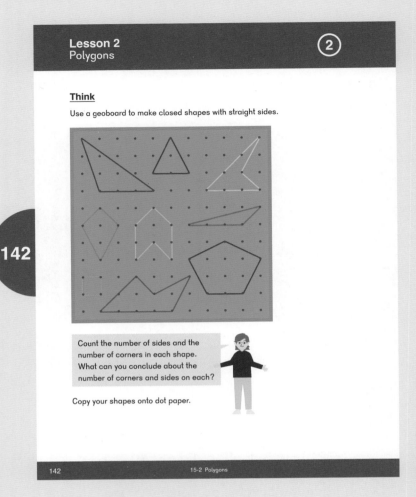

Think

Have students investigate and make closed shapes with straight sides on a geoboard. Shapes cannot overlap.

Have students copy the shapes they make onto Dot Paper (BLM).

Pose Emma's question about the lines and corners on the shapes.

Learn

Discuss the term "polygon."

Have students count to verify Dion's comment about the number of sides and corners.

Discuss the terms for specific polygons. A triangle has three sides and three corners. Students can relate the term "tri" to 3, as in tricycle, triplets, triathlon, or trio.

A quadrilateral has four sides and four corners. Students can relate the term "quad" to 4, as in a quad bike or quadruplets.

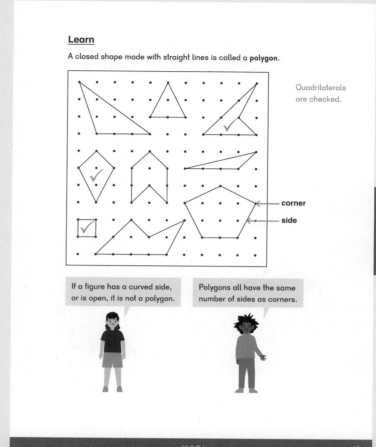

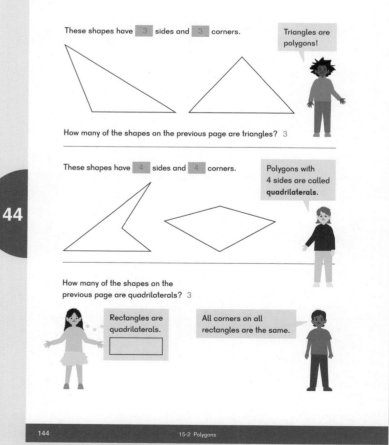

Do

1. Students should use pattern blocks and rotate them to see which shapes have sides of equal length.

2. Use Dot Paper (BLM).

3. Have students draw pentagons and hexagons on dot paper or identify any that they drew for 2.

4. While the term "concave" is not used with students, they should see that these are also corners:

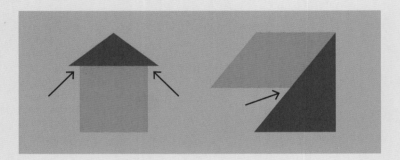

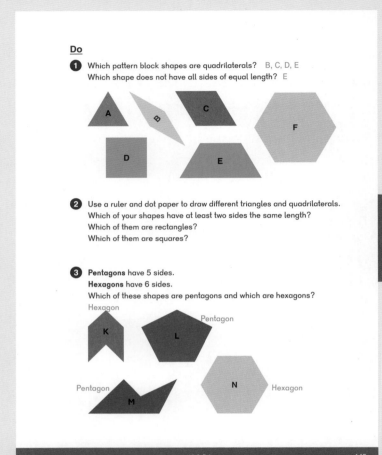

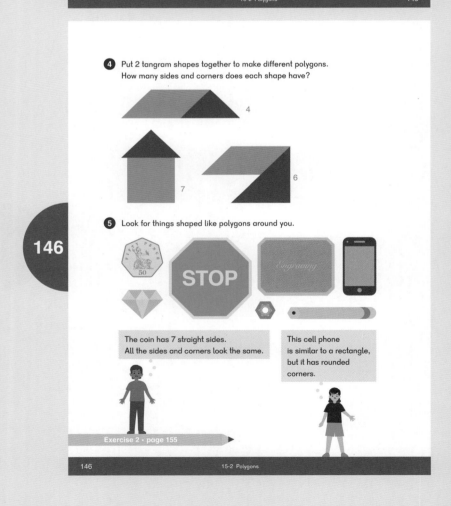

Activities

▲ Polygon Walk

Materials: Roll of painter's tape for each group

Have students work in groups of three. Start with the end of the tape at a point on the floor. Have Partner One walk 5 or fewer steps in a straight line in one direction and stop. Partners Two and Three will lay a piece of tape along the line that Partner One walked.

Partner One turns and starts a new line of walking, with her partners taping the line. She may make several lines but must not cross a line and must end at the same spot she started walking. This will make a polygon.

Have students rotate through each group's polygon, walking along the tape and counting the corners and sides of each polygon as they walk.

▲ Polygon Pictures

Materials: Paper, ruler, crayons or colored pencils

Have students use a ruler and draw random lines on a piece of paper.

Students then identify shapes on their paper and color them based on the number of sides.

For example, all triangles might be blue, all quadrilaterals green, all pentagons yellow, etc.

You can also try this activity with masking tape and chalk on the playground.

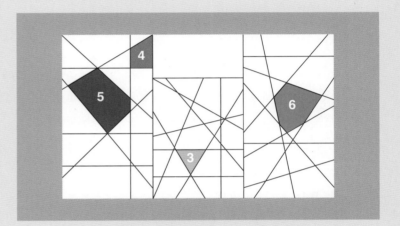

◀ **Exercise 2 • page 155**

Lesson 3 Semicircles and Quarter-circles

Objectives

- Identify semicircles and quarter-circles.
- Make new shapes by combining semicircles and quarter-circles.

Lesson Materials

- Semicircles (BLM) for each student

Or

- 2 paper circles/plates per student
- Paper squares (the radius of the circles should be the same length as the side of the square), 2 per student

Think

Provide students paper circles (either from Semicircles (BLM), paper plates, or circles cut from regular paper) and have them fold and cut one in half and another into quarters.

Discuss what students notice about the shapes they are making.

Learn

When discussing the shapes, ask students what they notice about the names "semicircle" and "quarter-circle." They could draw a correlation to fourths or even quarts.

Ask students, "Are these polygons?"

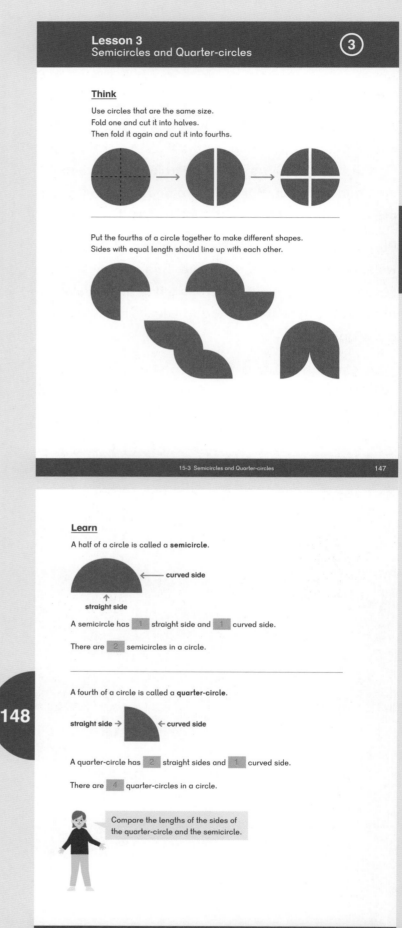

182　　　　　　　　　　　　　Teacher's Guide 2B Chapter 15　　　　　　　　　　© 2017 Singapore Math Inc.

Do

Students will use circles and squares from Semicircles (BLM). If not using the BLM, ensure that the length of one side on the square is the same as the length of one straight side on a quarter-circle (the radius).

Activity

▲ **Musical Shapes**

Materials: Large paper shapes (quadrilaterals, rectangles, circles, semicircles, quarter-circles, triangles, pentagons, and hexagons) taped to the floor, corresponding shape cards, music

Place shapes in large circle on the floor. Similar to Musical Chairs, have students walk around the circle until the music stops. At that point they stand on the nearest shape.

Pull a shape card, and all students on that shape are out. Continue to play rounds until only one student is left.

Exercise 3 • page 159

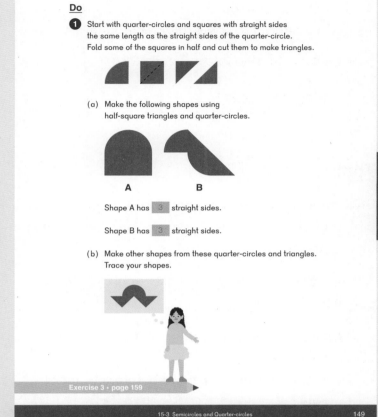

Lesson 4 Patterns

Objective
- Make and complete patterns with two-dimensional shapes according to one or two attributes and explain the patterns.

Lesson Materials
- Tangram shapes

Think

Give students tangram shapes. Have them select 3 or 4 shapes and put them in a row, and then continue their pattern.

Ask students to share their pattern. Ask students questions from **Think**:

- What changes in the pattern? (Is it the direction, color, shape?)
- How many shapes are there before the pattern repeats?
- What shape comes next?

Discuss Dion's tangram pattern.

Learn

Note that in Dion's pattern, some students might say that the direction the triangle is pointing changes. Students are being introduced to the concept of orientation. Tell them that when a shape is turned or rotated, it defines the shape's **orientation**.

Orientation doesn't change the shape. It is a rotation of the earlier shape. The triangles' orientation rotates, or turns:

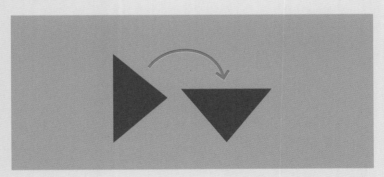

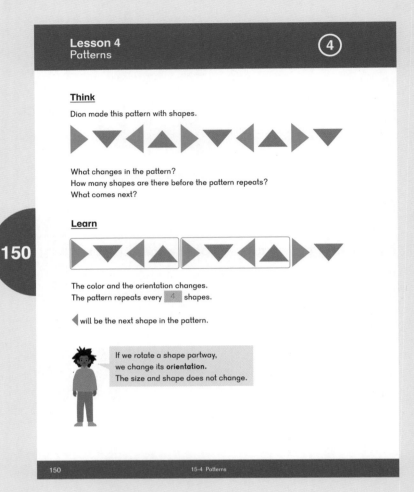

Do

The patterns involve shape, size, color, and orientation, but not necessarily all four in one pattern.

④ Students should see that if the 10th shape is a circle, so will the 100th be, as 100 is 10 tens.

They may also see that if the 5th is a square, the 15th will also be a square (as will the 25th, 35th, 75th, etc.).

Additionally, knowing that the 10th shape is a circle, students may know that the 20th, 30th, 70th, and 100th will also be circles.

(Odd and even numbers will be taught in **Dimensions Math® 3A**.)

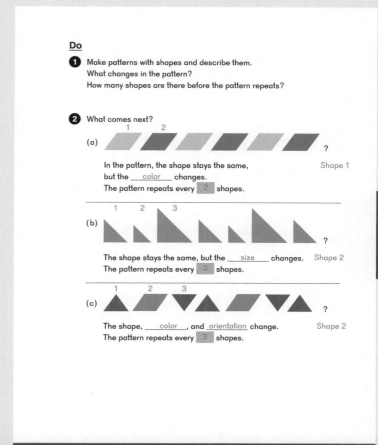

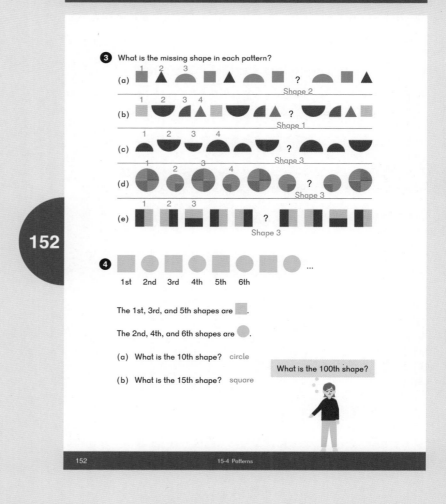

⑤ – ⑥ Students should be using the textbook to solve these problems. Students who are struggling could draw the continuing patterns if needed.

⑤ Note that students can find the 12th shape by:

(a) counting by twos.
(b) and (c) counting by threes.
(d) counting by fours.

To extend:

- Have students discuss how to find the shape in a specified position by division.
- Have students find the 13th shape (multiple of repeat + 1).

Activity

▲ Patterns

Materials: Pattern or attribute blocks

Have students work in pairs. Partner One should create a pattern. Partner Two can find the next three shapes to the pattern.

For a greater challenge, Partner One can create a pattern with missing pieces and Partner Two can find the missing pieces.

Exercise 4 • page 163

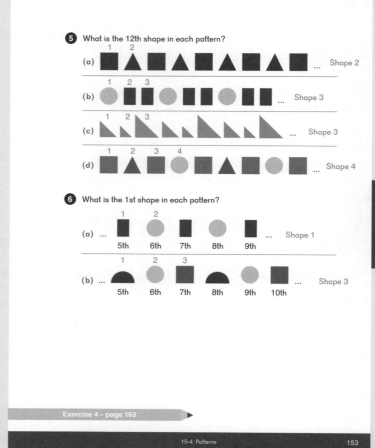

Lesson 5 Solid Shapes

Objective

- Identify attributes of three-dimensional shapes.

Lesson Materials

- Three-dimensional objects in the shape of cones, cylinders, spheres, cubes, and cuboids (or rectangular prisms) such as canisters, boxes, and frisbees
- Geometric solids shape sets

Think

Provide students with objects and have them sort the objects into groups. Students may group objects in many ways.

Examples include:

- Size
- Color
- Things that roll or slide
- Things with flat or curved sides
- Things that are shaped like a box and things that aren't

Have students explain their reasoning for the way they sorted and grouped the objects.

Repeat and ask students to look at the shapes of the objects and have them put similar shapes into groups.

Learn

The terms "face" and "edge" are introduced to help distinguish from the ambiguous "side."

Note that the objects in **Think** are similar to geometric shapes. The traffic cone, for example, has an additional base that is attached to the cone. A party hat is open on the bottom.

Have students count the number of sides and the number of corners on the cuboid and cube objects in **Think**. Remind students that with polygons, the number of sides and corners is the same. Ask if that is true of some three-dimensional solids.

Faces meet at edges. A sphere has only one curved surface. It does not have any edges.

On a three-dimensional object, when 2 faces come together, they make one edge.

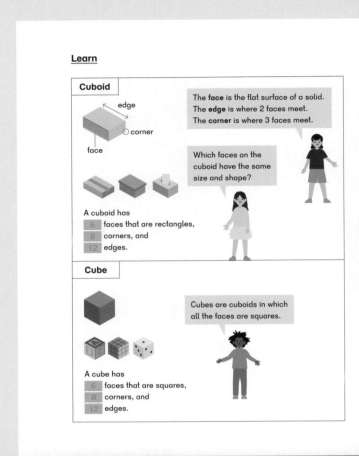

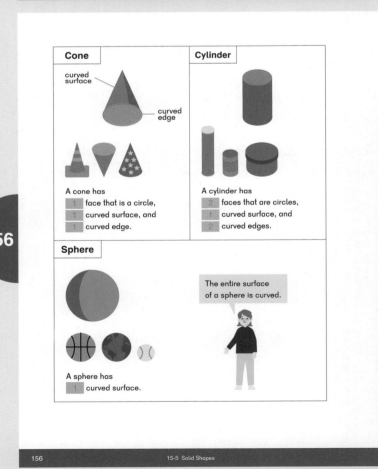

Do

Students should use the objects from **Think** or the same 5 shapes from a geometric solids shape set.

2 – 3 Have students use the solid shapes from geometric solids sets.

Activity

▲ Connect 4 Solids

Materials: Connect 4 Solids (BLM), two-color counters, paper clips

Use a pencil and paper clip as the spinner.

Players take turns spinning the spinner to determine the shape to cover up. If a player's spin lands on a shape that is already taken, he should remove his opponent's counter and replace it with his own.

The first player to get 4 counters in a row is the winner.

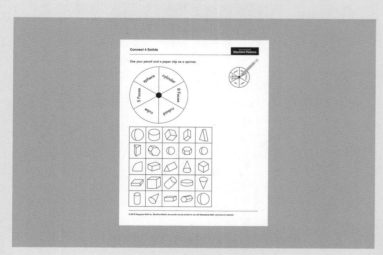

Exercise 5 • page 167

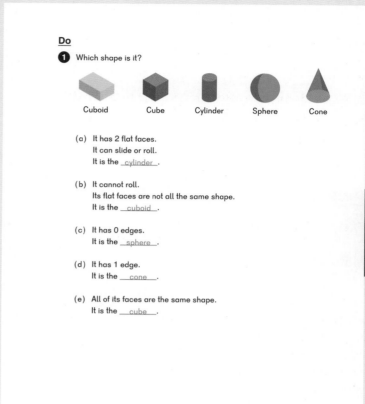

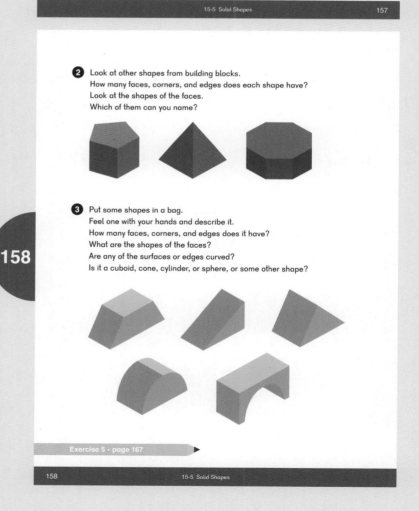

Lesson 6 Practice

Objective
- Practice concepts from the chapter.

After students complete **Practice** in the textbook, have them continue to practice concepts with activities from the chapter.

Activity

▲ Block-it Game

Materials: Pattern blocks

Each partner receives three each of the following pattern blocks, worth the following point values:

Green triangle	3 points
Blue rhombus	4 points
Red trapezoid	5 points
Yellow hexagon	6 points

The game begins with one yellow hexagon starting block placed on the playing surface. This piece does not belong to either player.

The scoring for each play is the sum of the values of the block placed and those that it touches on a side. Play continues until both players use all of their pieces.

For example:

Player One places one of her blocks so that one side of the block is touching the starting hexagon along its entire side on the playing surface. This move is worth 11 points.

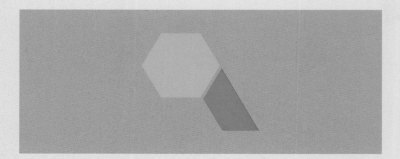

Player Two adds a trapezoid that touches Player One's trapezoid and the hexagon. This play is worth 16 points.

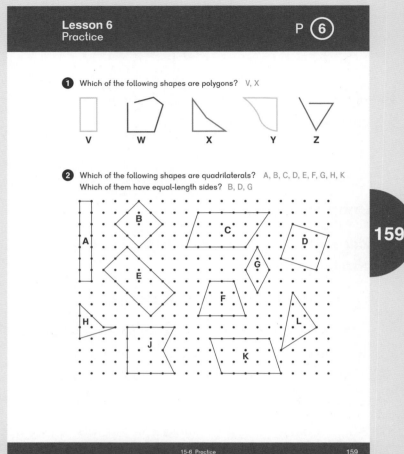

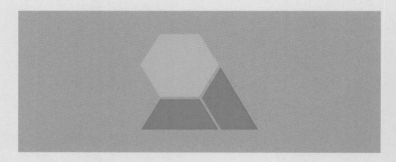

The player with the most total points after all pieces have been used is the winner.

Exercise 6 • page 171

Brain Works

★ **Patterns**

Look at the following patterns.

How many dots will be in the 4th group?

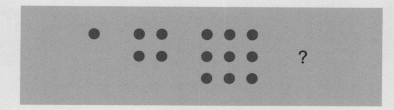

How many dots will be in the 5th group?

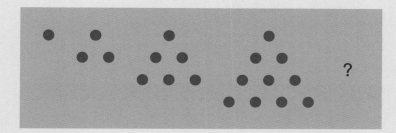

More squares are added to the right column following the same pattern until the last column is 9 squares tall. At that point, how many squares are there in all?

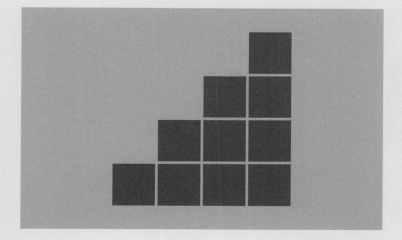

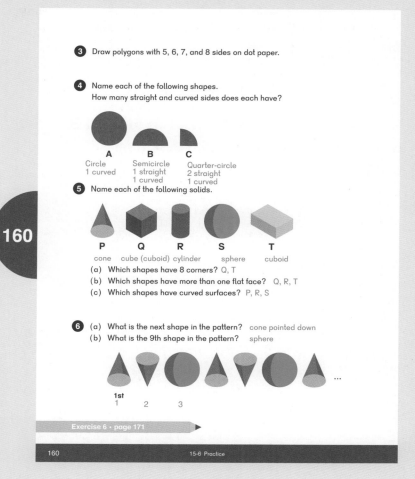

Review 4

Objective

- Review content from Chapter 1 through Chapter 15.

Two reviews are provided for students to recall and practice content from both **Dimensions Math® 2A** and **2B**.

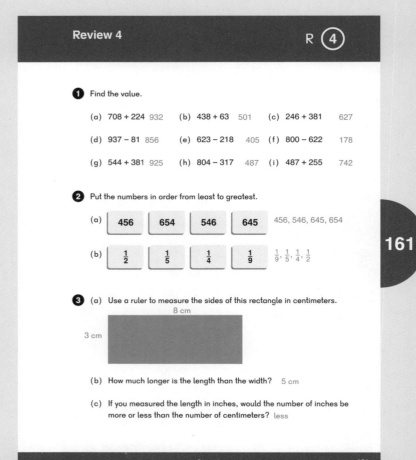

Exercise 7 • page 175

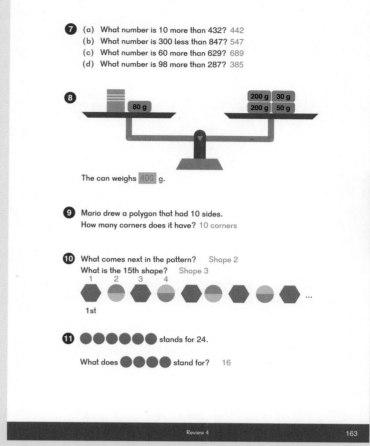

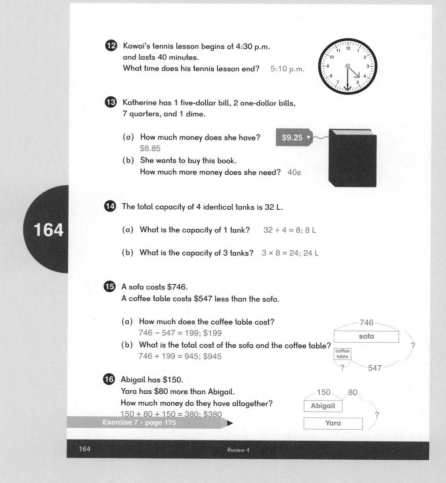

Review 5

Objective

- Review content from Chapter 1 through Chapter 15.

Review 5 R ⑤

① Find the value.

(a) 738 + 9   747   (b) 870 + 40   910   (c) 611 + 99   710

(d) 872 − 4   868   (e) 310 − 20   290   (f) 800 − 98   702

(g) 544 + 98   642   (h) 100 − 72   28   (i) 427 − 52   375

② (a) 246 + 70 = 316   (b) 185 − 7 = 178

(c) 333 + 98 = 431   (d) 565 − 40 = 525

③

A   B

(a) How many squares were used to make each rectangle? Write a multiplication equation for each.   Possible answer: A: 3 × 7 = 21; B: 4 × 3 = 12

(b) Each square used to make the figures above has a side 2 cm long. How long and wide is rectangle A?   14 cm long, 6 cm wide

(c) Rectangle A and B are joined to make a larger rectangle. How long and wide is the new rectangle?   22 cm long, 6 cm wide

165

---

④ Find the value.

(a) $8.20 + 65¢ $8.85   (b) 75¢ + 85¢ $1.60

(c) $2.87 + $4 $6.87   (d) $5.90 + $3.15 $9.05

(e) $1 − 20¢   80¢   (f) $5.05 − 55¢ $4.50

(g) $6.00 − $1.35 $4.65   (h) $3.25 − $2.35 $0.90

⑤ What time will it be…

(a) 25 minutes after 3:45 p.m.?   4:10 p.m.

(b) 3 hours after 10:20 a.m.?   1:20 p.m.

⑥ How many straight sides does each shape have?

(a) 2   (b) 0

(c) 8   (d) 2

166

---

194   Teacher's Guide 2B Chapter 15   © 2017 Singapore Math Inc.

Exercise 8 • page 181

7 This picture graph shows the number of items sold at a bake sale.

Items Sold				
■				
■				■
■			■	■
■	■		■	■
■	■		■	■
■	■	■	■	■
Cookies	Muffins	Scones	Cereal Bars	Krispie Treats

Each ■ stands for 5 items.

(a) Which item was sold the most? Cookies
(b) Which item was sold the least? Scones
(c) Which 3 items were the most popular? Cookies, Cereal Bars, Krispie Treats
(d) How many Krispie Treats were sold? 5 × 5 = 25; 25 Krispie Treats
(e) Each Cereal Bar was sold for $2.
    How much money was made from selling the Cereal Bars? 4 × 5 = 20
(f) The Scones sold for $30.   20 × 2 = 40;
    How much did each Scone cost? 30 ÷ 5 = 6; $6 $40

8 Don painted $\frac{3}{5}$ of a fence.
What fraction of the fence still needs to be painted? $\frac{2}{5}$

9 Kai ate $\frac{1}{3}$ of a pizza.
Isabella ate $\frac{1}{4}$ of the pizza.
Who ate more pizza? Kai

10 Matt wants to save $95 to buy a bike.
He has saved $5 every week for 9 weeks.

(a) How much more does he have to save?
    9 × 5 = 45; 95 − 45 = 50; $50
(b) If he continues to save $5 a week,
    how many more weeks will it take to save the full amount?
    50 ÷ 5 = 10; 10 more weeks

11 The Olympic Class ferry has a capacity of 144 cars,
the Super Class ferry has a capacity of 188 cars,
and the Jumbo Class ferry has a capacity of 200 cars.

(a) How many more cars can the Jumbo Class
    ferry carry than the Olympic Class ferry?
    200 − 144 = 56; 56 cars
(b) How many cars can all 3 types of ferries carry?
    144 + 188 + 200 = 532; 532 cars

12 Matias has the same number of dimes as nickels.
He has 60¢ altogether.
How many of each coin does he have?
4 dimes, 4 nickels

Exercise 1 • pages 153–154

Chapter 15 Shapes

Exercise 1

Basics

1. Check the figures that are made from only curved lines.

2. Check the figures that are made from only straight lines.

3. Check the figures that are closed shapes.

Practice

4. Use a ruler to draw only straight lines to make each of these figures closed shapes.

   (a)   (b)   (c)

   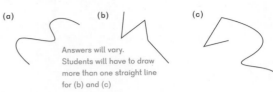

   Answers will vary. Students will have to draw more than one straight line for (b) and (c).

5. Draw only curved lines to make each of these figures closed shapes.
   Answers will vary.

   (a)   (b)   (c)

   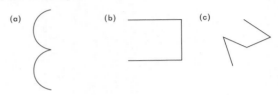

6. Draw a closed shape that has 4 straight sides and at least 1 curved side.
   Answers will vary.

Exercise 2 • pages 155–158

Exercise 2

Basics

1. Check the figures that are polygons.

   [✓] [ ] [ ] [✓]

2. Check the figures that are quadrilaterals.

   [✓] [✓] [ ] [ ]

3. (a) A hexagon has __6__ sides and __6__ corners.

   (b) A pentagon has __5__ sides and __5__ corners.

   (c) A polygon with 20 sides has __20__ corners.

   (d) A polygon has __0__ curved sides.

   (e) The fewest number of sides a polygon can have is __3__.

Practice

4. Use the dot paper and a ruler to draw two of each of the following shapes.

   A Triangle
   B Quadrilateral
   C Square
   D Pentagon
   E Hexagon

   Answers will vary.

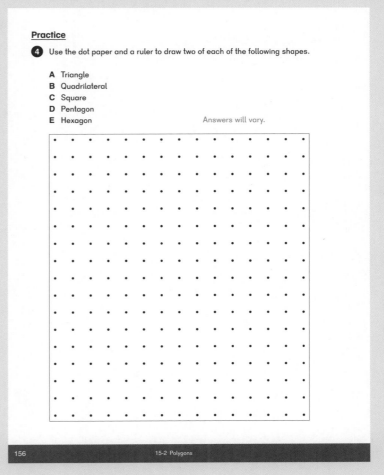

5. Draw straight lines on each figure to divide it into the given shapes.

   Answers will vary.

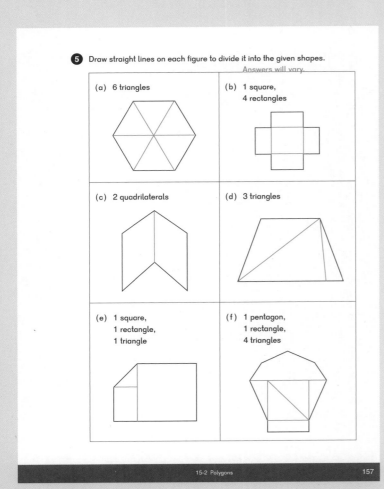

(a) 6 triangles
(b) 1 square, 4 rectangles
(c) 2 quadrilaterals
(d) 3 triangles
(e) 1 square, 1 rectangle, 1 triangle
(f) 1 pentagon, 1 rectangle, 4 triangles

Challenge

6. Trace and cut out two copies of each of the shapes below.

   Draw straight lines to show how each of these figures could be formed from these shapes.
   Use your cut outs to help you.   Answers will vary.

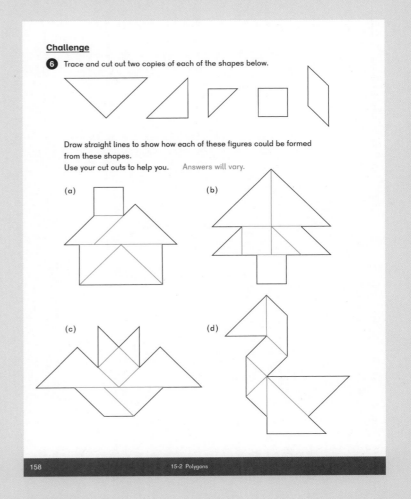

Exercise 3 • pages 159–162

Exercise 3

Basics

1. Use a ruler to draw a straight line to divide each circle into semicircles.

   Answers may vary.

   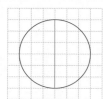

2. Use a ruler to draw straight lines to divide each circle into quarter-circles.

   Answers may vary.

   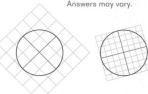

3. (a) A semicircle has __1__ curved side and __1__ straight side.

   (b) There are __2__ semicircles in a circle.

   (c) A quarter-circle has __1__ curved side and __2__ straight sides.

   (d) There are __4__ quarter-circles in a circle.

Practice

4. Check the figures that are semicircles.

   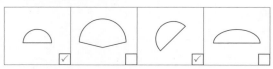

5. Check the figures that are quarter-circles.

   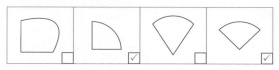

6. Draw lines to divide each figure into semicircles and quarter-circles.

   (a)   (b)

   (c)    (d)

7. Draw straight lines on each figure to divide it into the given shapes.

   (a) 1 semicircle,
   1 rectangle,
   1 triangle

   (b) 1 square,
   1 triangle,
   3 quarter-circles

   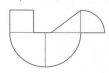

8. Fill in the blanks with the names of the shapes:
   quarter-circle, semicircle, triangle, or rectangle.

   (a) This figure is made from

   2 __quarter-circles__,

   a __triangle__,

   and a __rectangle__.

   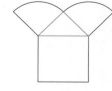

   (b) This figure is made from

   a __semicircle__,

   a __triangle__,

   and a __rectangle__.

Challenge

9. How many quarter-circles are shown in each drawing?

   (a) 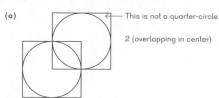 ← This is not a quarter-circle.

   2 (overlapping in center)

   (b) 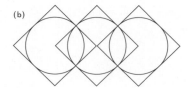 6

   (c) 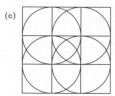 16 (4 overlapping quarter-circles in center, 4 pairs with 2 overlapping on top, bottom, left and right, 4 more on corners.)

Teacher's Guide 2B Chapter 15 © 2017 Singapore Math Inc.

Exercise 4 • pages 163–166

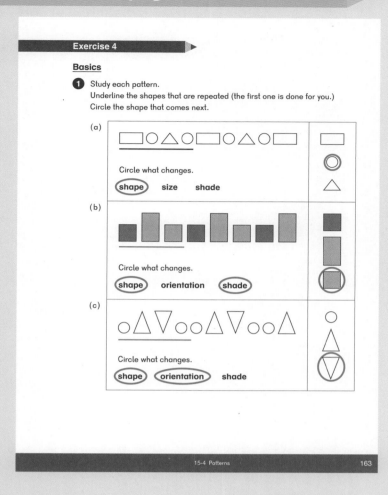

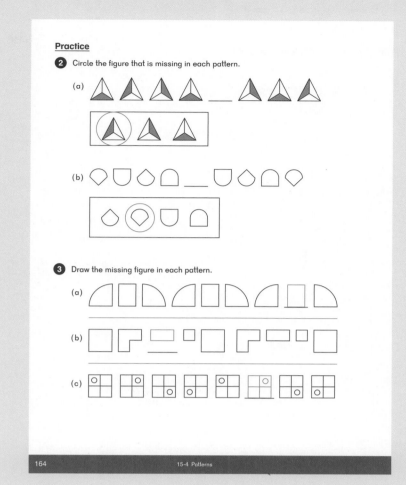

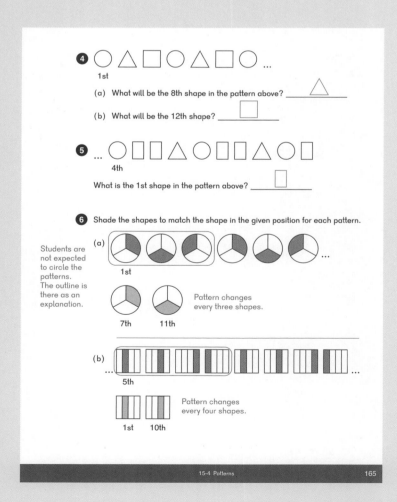

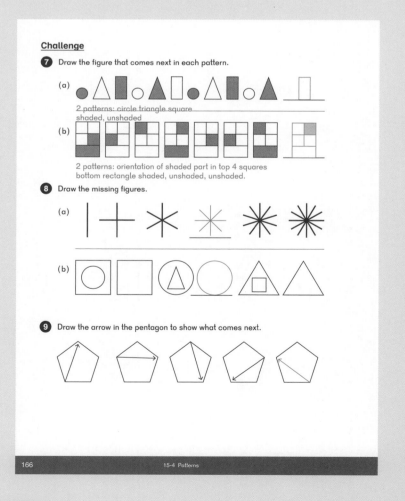

Exercise 5 • pages 167–170

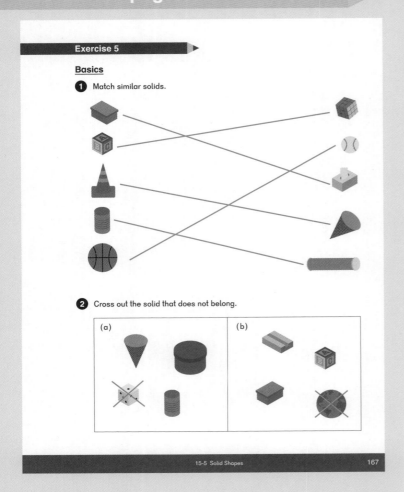
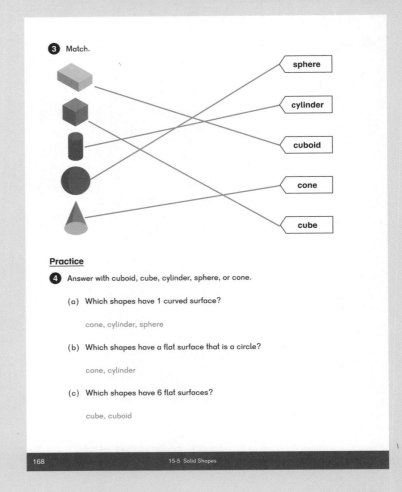
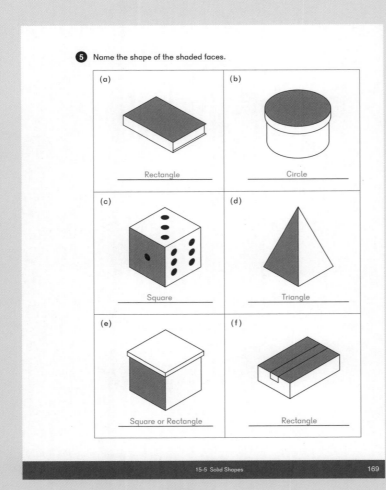
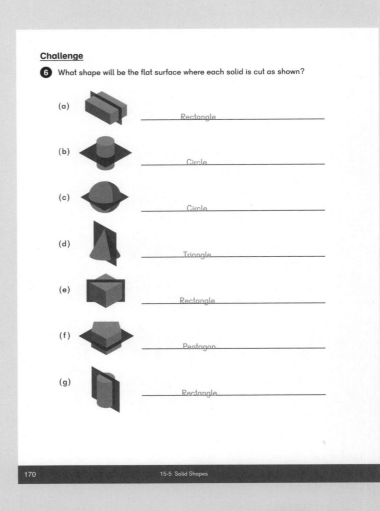

Exercise 6 • pages 171–174

Exercise 6

Check

1. Write the number of straight lines in each of the following figures.
   Circle the figures that are closed.
   Cross out the polygon.

   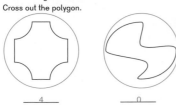
      4   0

   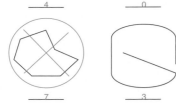
      7   3

2. Draw straight lines to divide each figure into the fewest possible quarter-circles, half-circles, quadrilaterals, and triangles.

   (a)   (b)   (c)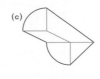

3. Write the name of the shape that comes next in each pattern.

   (a)  Pentagon

   (b)  Sphere

4. Draw the missing figure in the pattern.

5. Complete the table.

Solid	Number of flat faces	Number of curved surfaces	Number of corners
Cuboid	6	0	8
Cone	1	1	1
Cylinder	2	1	0
Sphere	0	1	0
	7	1	12
	3	1	4

6. Use a ruler to copy the figures onto the dot grid. Then draw straight lines to divide each of your copied figures into triangles.

   Answers will vary. Example are shown.

   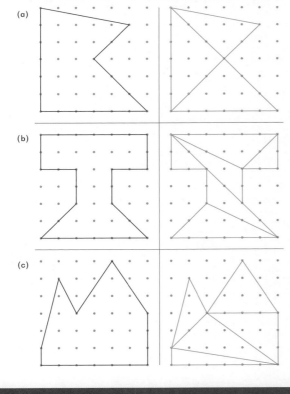

Challenge

7. A paper roll cardboard is shaped like a ___cylinder___.

   If it is cut along the dotted line and flattened out, what flat shape will it be? ___Rectangle___

8. How many triangles are in this shape?
   Hint: There are more than 4.

   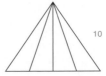 10

9. (a) Draw the arrow in the figure to show what comes next.

   1st   The base of the arrow rotates two corners clockwise each time.

   (b) Draw the arrow in the figure to show the 10th figure if the pattern above is continued.

     The pattern will repeat after the 7th shape.

Exercise 7 • pages 175–180

Exercise 7

Check

1 680 829 983 806 68

Use the numbers above for each of the following problems.

(a) The numbers with 8 in the tens place are __680, 983__.

(b) The numbers with 8 in the hundreds place are __829, 806__.

(c) The digit in the hundreds place in 893 is __5__ more than the digit in the ones place.

(d) The digit in the ones place in __68__ is 2 more than the digit in the ones place in __806__.

(e) Write the numbers in order from least to greatest.

| 68 | 680 | 806 | 829 | 983 |

(f) 30 more than __829__ is 859.

(g) 200 less than __982__ is 782.

(h) 70 more than __680__ is 750.

(i) 6 less than __983__ is 977.

(j) The digits in __983__ add to 20.

2 Write the missing digits.

(a) 2 7 **5**
   + 3 8 6
   ─────
     6 6 1

(b) 9 3 0
   −   2 7 8
   ─────
     6 5 2

3 What fraction of each shape is shaded and what is the name of the shape?

(a) $\frac{2}{3}$ of the __semicircle__ is shaded.

(b) $\frac{3}{5}$ of the __pentagon__ is shaded.

4 (a) The time is __17__ minutes past __10__.

(b) The time is __25__ minutes to __1__.

5 How much weight needs to be added to balance the scale?

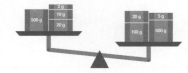

__193__ g needs to be added to the left side.

6 Use a centimeter ruler to measure all three sides of this triangle. The total length of all three sides is __21__ cm.

7 What is the total amount of money?

$ __7__ . __98__

8 5 equal size buckets can hold 40 L of water.

(a) How much water can 1 of these buckets hold?

40 ÷ 5 = 8
One bucket can hold 8 L.

(b) 3 buckets are needed to fill a barrel. What is the capacity of the barrel?

3 × 8 = 24
The capacity is 24 L.

9 There were 389 adults and 562 children at an amusement park. How many fewer adults were there than children?

562 − 389 = 173
There are 173 fewer adults than children.

10 Chapa saved $3.85.
Isabella saved $5.35.

(a) How much more did Isabella save than Chapa?
$5.35 − $3.85 = $1.50   Isabella saved $1.50 more than Chapa.

(b) How much did they save altogether?

$5.35 + $3.85 = $9.20
They saved $9.20 altogether.

11 A notebook cost $2.45.
A binder costs 5 quarters and 1 dime more than the notebook. How much does the binder cost?

$2.45 + $1.35 = $3.80
The binder costs $3.80.

12 Eliza ate $\frac{1}{3}$ of a pizza.
Emiliano ate $\frac{1}{5}$ of the pizza.
Who ate more of the pizza?

Eliza ate more pizza.

13. This picture graph shows the number of some animals in a pet store.

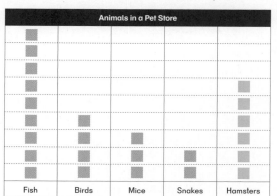

There are 15 mice and snakes altogether.

(a) Each ■ stands for __3__ animals.   15 ÷ 5 = 3

(b) How many of each animal are there?

   Fish __27__   Birds __12__   Mice __9__   Snakes __6__

(c) There are 18 hamsters. Draw the correct number of ■ for the hamsters.

(d) The pet store got more fish and now has twice as many fish.
   How many fish are there now? __54__   27 + 27 = 54

(e) How many of all 5 kinds of animals are there altogether now? __99__
   54 + 12 + 9 + 6 + 18 = 99

Challenge

14.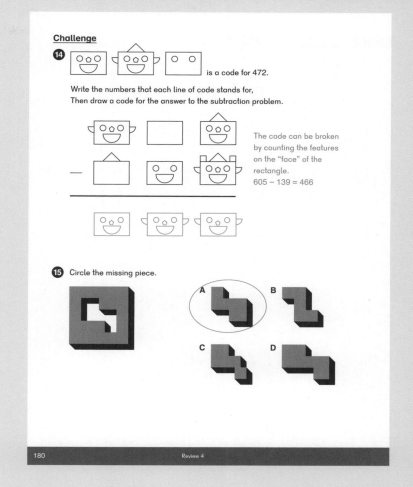

is a code for 472.

Write the numbers that each line of code stands for,
Then draw a code for the answer to the subtraction problem.

The code can be broken by counting the features on the "face" of the rectangle.
605 − 139 = 466

15. Circle the missing piece.

Exercise 8 • pages 181–186

Exercise 8

Check

1. Write >, <, or = in the ◯.

   (a) 500 + 40 + 3 ◯(>) 30 + 400 + 5
   (b) 14 + 36 + 25 ◯(<) 27 + 29 + 22
   (c) 580 + 70 ◯(=) 710 − 60
   (d) $8.25 − $1.80 ◯(<) $6.25 + $1.80
   (e) $4.30 ◯(>) 403¢
   (f) $\frac{1}{5}$ ◯(>) $\frac{1}{8}$
   (g) 7 × 4 ◯(>) 3 × 9
   (h) 35 ÷ 5 ◯(<) 32 ÷ 4
   (i) 1 m ◯(=) 100 cm
   (j) 510 g + 480 g ◯(<) 1 kg

2. (a) 66 + [34] = 100
   (b) [583] − 99 = 484
   (c) [314] + 8 = 322
   (d) 300 − 64 = [236]
   (e) 45 ÷ [5] = 9
   (f) 4 × [7] = 28
   (g) 62 + 24 + 30 + 15 = [228] − 97

3. (a) $\frac{3}{8}$ and $\frac{5}{8}$ make 1.
   (b) $\frac{5}{12}$ and $\frac{7}{12}$ make 1.

4. How many rectangular faces are on each of these solids?

   (a)   (b)   (c)

      5   1   3

5. This clock shows the time Alex began doing his chores on Saturday.
   Write the times using a.m. or p.m.

   (a) What time did he begin his chores? 10:50 a.m.
   (b) He finished his chores 35 minutes later.
       What time did he finish his chores? 11:25 a.m.
   (c) Then, he went outside to play for 4 hours.
       What time did he finish playing? 3:25 p.m.

6.

   ...   7th   ...

   Color the shape to show the shape in the given position in the pattern.

   1st   3rd   12th   18th

7. A tank contains 132 L of water.
   389 L are needed to fill it.
   What is the capacity of the tank?

   132 + 389 = 521
   The capacity of the tank is 521 L.

8. A bag of rice weighs 4 kg.
   The price for 1 bag is $5.
   Mr. Baker bought some rice for $25.

   (a) How many bags of rice did he buy?
       25 ÷ 5 = 5   He bought 5 bags of rice.
   (b) How many kilograms of rice does he have?

   4 × 5 = 20
   He has 20 kg of rice.

9. Jasmine cut a cake into 10 equal pieces.
   She and her 2 friends each ate a piece.

   (a) What fraction of the cake did they eat?
       $\frac{3}{10}$
   (b) What fraction of the cake is left?
       $\frac{7}{10}$

10. Circle the figure that completes the design.

11. Use the information below to find the missing lengths.

         12 in   20 in

    (a)   (b)   (c)
       3 in   4 in   14 in

    10 in.
    (d)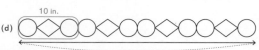

    Students are not expected to circle the patterns. The outline is there as an explanation.   40 in

    (e) If 2 more shapes are added to continue the pattern in (d) above, what would be the new length?

       47 in

12. (a) Use the information below to complete the bar graph.

    $2.25 in quarters.
    $1.70 in dimes.
    The same number of nickels as dimes.
    12 pennies.

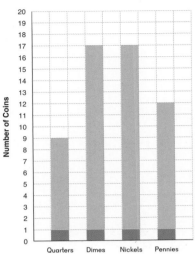

(b) The total number of coins is __55__.

(c) The nickels make $__0.85__

Challenge

13. Pablo cut a pie into fourths and put one piece on a plate.
    He then cut that piece into thirds and ate one of those pieces.
    What fraction of the pie is left?
    Hint: Draw a picture.

    $\frac{11}{12}$ of the pie is left.

    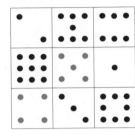

    Students do not have to make a precise drawing. From the drawing, one of the fourths is cut into 3 pieces, so if all are cut into 3 pieces there would be 12. One piece is eaten.

14. Use the digits 0, 1, 2, 4, 5, 7, 8, and 9 to fill in the boxes.
    Each number can be used once.

    $\boxed{4} \times \boxed{5} = \boxed{2}\boxed{0}$

    $\boxed{8} + \boxed{9} = \boxed{1}\boxed{7}$

15. Draw the correct number of missing dots in the blank squares to complete the puzzle.

    Arrangement of dots not important. Sum of dots vertically, horizontally, and diagonally is 15.

Blackline Masters for 2B

All Blackline Masters used in the guide can be downloaded from dimensionsmath.com.
This lists BLMs used in the **Think** and **Learn** sections.
BLMs used in **Activities** are included in the Materials list within each chapter.

Array Dot Cards – 3	**Chapter 9:** Lesson 1
Array Dot Cards – 4	**Chapter 9:** Lesson 5
Bar Graph 14-2	**Chapter 14:** Lesson 2
Clock Face	**Chapter 12:** Lesson 2
Dot Paper	**Chapter 11:** Lesson 3 **Chapter 15:** Lesson 2
Double Hundred Chart	**Chapter 8:** Lesson 4
Fraction Exercises	**Chapter 11:** Lesson 1
Hundred Chart	**Chapter 8:** Lesson 3
Multiplication Chart 9-1	**Chapter 9:** Lesson 1
Multiplication Chart 9-5	**Chapter 9:** Lesson 5
Number Cards	**Chapter 8:** Lesson 3
Semicircles	**Chapter 15:** Lesson 3